Kawasaki 900 and 1000 Fours Owners Workshop Manual

by George Collett

With an additional Chapter on the Z900 and Z1000 models

by Pete Shoemark

Models covered:

Z1 Series.	903 cc.	UK March 1973 to October 1976
Z1 Series.	903 cc.	USA 1973 to 1975
Z900.	903 cc.	UK January 1976 to October 1976
KZ900.	903 cc.	USA 1975 to 1977
Z1000.	1015 cc.	UK October 1976 to 1977
KZ1000.	1015 cc.	USA 1976 to 1977

ISBN 978 1 85010 623 4

© Haynes Group Limited 2013

(222-2AM7)

Haynes Group Limited
Haynes North America, Inc

www.haynes.com

Acknowledgements

The author wishes to thank Kawasaki Motors (UK) Limited for the technical assistance given so freely whilst this manual was being prepared, and for permission to reproduce their drawings. The model used for the photographic sequences was supplied by Davick Motique of Long Eaton Nottinghamshire, who also provided the new parts and gaskets used in the rebuild. Brian Horsfall assisted with the dismantling and rebuilding of the machine and also devised various ingenious methods for overcoming the lack of service tools. Les Brazier arranged and took all the photographs. The author would also like to thank Jeff Clew for his guidance and editorship.

Finally thanks to NGK Spark Plugs (UK) Limited for the provision of spark plug photographs, Avon Rubber Company Limited for their advice on tyre fitting and to the Bristol Kawasaki Centre for permitting us to photograph the Z1000 model featured on the front cover.

About this manual

The author of this manual has the conviction that the only way in which a meaningful and easy to follow text can be written is to first do the work himself, under conditions similar to those found in the average home. As a result, the hands seen in the photographs are sometimes the hands of the author together with another engineer who assisted. The machine photographed was a used model that had covered three thousand miles, so that the conditions encountered would be similar to those found by the average rider.

Unless specially mentioned, and therefore considered essential, Kawasaki service tools have not been used. There is invariably some alternative means of slackening or removing some vital component when service tools are not available and the risk of damage has to be avoided at all costs.

Each of the seven Chapters is divided into numbered Sections. Within the Sections are numbered paragraphs. In consequence, cross reference throughout this manual is both straightforward and logical. When a reference is made 'See Section 1.6' it means see Section 1, paragraph 6 in the same Chapter. If another Chapter were meant, the text would read 'See Chapter 4, Section 1.6'. All the photographs are captioned with a Section paragraph number to which they refer and are always relevant to the Chapter text adjacent.

Figure numbers (usually line illustrations) appear in numerical order, within a given Chapter. Figure 1.2 therefore refers to the first figure in Chapter 1.

Left hand and right hand descriptions of parts of the machine or the machine itself, refer to the right and left side of the machine, with the rider seated in the normal riding position.

Motorcycle manufacturers continually make changes to specifications and recommendations, and these, when notified, are incorporated into our manuals at the earliest opportunity.

Whilst every care is taken to ensure that the information in this manual is correct, no liability can be accepted by the authors or publishers for loss, damage or injury, caused by any errors in or omissions from the information given.

Contents

Chapter	Section	Page
Introductory pages	Dimension and weights	4
	Introduction to the Z1 series	5
	Ordering spare parts	5
	Safety first!	6
	Routine maintenance	7
	Quick glance maintenance	9
	Lubrication points and Recommended lubricants	10
Chapter 1/Engine, clutch and gearbox	Engine/gearbox removal	13
	Examination and renovation - general	36
	Dismantling - general	18
	Reassembly - general	42
	Starting and running a rebuilt unit	56
Chapter 2/Fuel system and lubrication	Air cleaner	78
	Carburettors	69 - 78
	Oil pump	79
	Petrol tank and tap	69
	Rear chain lubrication	79
Chapter 3/Ignition system	Condensers - removal and replacement	85
	Contact breakers - adjustment	85
	Crankshaft alternator - checking output	84
	Ignition coils - checking	85
	Ignition timing	85 - 87
	Spark plugs - checking and resetting the gaps	87
Chapter 4/Frame and forks	Frame - examination and renovation	89
	Front forks	88 - 89
	Steering head bearings	89
	Suspension units - rear	97
	Swinging arm fork	97 - 133
Chapter 5/Wheels, brakes and tyres	Brakes	
	Front	110
	Rear	106
	Cush drive: examination and renovation	109
	Final drive chain: examination and lubrication	109
	Tyres: removal and replacement	116
	Wheels	
	Front	102 - 106
	Rear	106
Chapter 6/Electrical system	Alternator - crankshaft	120
	Battery - maintenance and charging procedure	120
	Fuse location	122
	Handlebar switches	126
	Silicon rectifier - location and replacement	120
	Voltage regulator - testing	121
	Wiring diagrams	129 - 130
Chapter 7/The Z900 and Z1000 models	General description	133
	Final drive chain - maintenance and renewal	133
	Hydraulic disc brakes - modifications	135
	Rear brake caliper: maintenance, removal and replacement	137
	Rear brake master cylinder: maintenance, removal and renovation	135
	Swinging arm fork: removal and renovation	133
	Wiring diagrams	129 - 130, 140 - 145
Metric conversion tables		147 - 148
Index		150
English/American terminology		149

General descriptions and specifications are given in each Chapter immediately after the list of Contents
Fault diagnosis is given at the end of each appropriate Chapter

Right-hand view of the Kawasaki 900cc series Z1

Model dimensions

Overall length	86.8 in (2.205 mm) US models 88.5 in (2.250 mm) European models
Overall width	31.5 in (800 mm) US models 32.3 in (820 mm) European models
Overall height	45.3 in (1.150 mm) US models 46.3 in (1.175 mm) European models
Wheelbase	58.7 in (1.490 mm)
Ground clearance	6.3 in (160 mm)
Dry weight	506 lb (230 kg)

Introduction to the Kawasaki Z1 series

When the 900 cc Z1 model was first introduced in 1972 it was obvious that Kawasaki had scored a huge success. The growth of the company had been little short of phenomenal, perhaps causing some people to wonder how it was achieved. The answer lay in the vast resources of the firm and the extent of their technological know-how, which extended into railroad, shipping, and aircraft transportation on a grand scale. All these activities rolled into one form Kawasaki Heavy Industries, a giant manufacturing complex that produces an astonishing variety of products and markets them all over the world.

This is just a bare outline of the industrial might of Kawasaki. It is the Motorcycle Division of Kawasaki Heavy Industries that has attracted most interest, the meteoric rise of this section being equal to that of the parent company being as a whole. In just a few years Kawasaki has become the fourth largest motorcycle manufacturer in the world and that in itself is quite an accomplishment when it is recalled that some European companies have been manufacturing machines for over 60 years. Kawasaki has now become seriously involved with racing, to such an extent that they field factory teams in trials, road racing and motocross, and were the only Japanese manufacturer to participate in the 48th I.S.D.T. held in Berkshire, U.S.A. More important is the readiness of Kawasaki to incorporate the hard learned lessons learnt at the race track, into their road going machines. In this way they have successfully capitalised on their competition successes, putting the knowledge they have gained at the disposal of all those who purchase their high quality products.

Ordering spare parts

When ordering spare parts for a Kawasaki it is advisable to deal direct with an official Kawasaki agent who should be able to supply most of the parts from stock. Parts cannot be obtained direct from Kawasaki U.K.; all orders must be routed via an approved agent as is common with most other makes.

Always quote the frame and engine numbers in full. The frame number is stamped on the left hand side of the steering head and the engine number on top of the crankcase to the rear of the cylinder block, on the right hand side.

It is always best to quote the colour scheme for any of the cycle parts that have to be ordered. Use only genuine Kawasaki parts. Pattern parts should be avoided as they are usually inferior in quality. Some of the more expendable parts such as bulbs, spark plugs, chains, tyres, oils and greases etc., can be obtained from accessory stores and motor factors, who have convenient opening hours, charge lower prices and can often be found nearer home. It is also possible to obtain parts on a Mail Order basis from a number of specialists who advertise regularly in the motorcycle magazines.

Frame number location

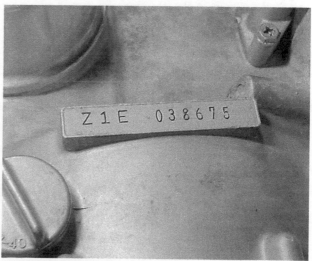

Engine number location

Safety first!

Professional motor mechanics are trained in safe working procedures. However enthusiastic you may be about getting on with the job in hand, do take the time to ensure that your safety is not put at risk. A moment's lack of attention can result in an accident, as can failure to observe certain elementary precautions.

There will always be new ways of having accidents, and the following points do not pretend to be a comprehensive list of all dangers; they are intended rather to make you aware of the risks and to encourage a safety-conscious approach to all work you carry out on your vehicle.

Essential DOs and DON'Ts

DON'T start the engine without first ascertaining that the transmission is in neutral.

DON'T suddenly remove the filler cap from a hot cooling system – cover it with a cloth and release the pressure gradually first, or you may get scalded by escaping coolant.

DON'T attempt to drain oil until you are sure it has cooled sufficiently to avoid scalding you.

DON'T grasp any part of the engine, exhaust or silencer without first ascertaining that it is sufficiently cool to avoid burning you.

DON'T allow brake fluid or antifreeze to contact the machine's paintwork or plastic components.

DON'T syphon toxic liquids such as fuel, brake fluid or antifreeze by mouth, or allow them to remain on your skin.

DON'T inhale dust – it may be injurious to health (see *Asbestos* heading).

DON'T allow any spilt oil or grease to remain on the floor – wipe it up straight away, before someone slips on it.

DON'T use ill-fitting spanners or other tools which may slip and cause injury.

DON'T attempt to lift a heavy component which may be beyond your capability – get assistance.

DON'T rush to finish a job, or take unverified short cuts.

DON'T allow children or animals in or around an unattended vehicle.

DON'T inflate a tyre to a pressure above the recommended maximum. Apart from overstressing the carcase and wheel rim, in extreme cases the tyre may blow off forcibly.

DO ensure that the machine is supported securely at all times. This is especially important when the machine is blocked up to aid wheel or fork removal.

DO take care when attempting to slacken a stubborn nut or bolt. It is generally better to pull on a spanner, rather than push, so that if slippage occurs you fall away from the machine rather than on to it.

DO wear eye protection when using power tools such as drill, sander, bench grinder etc.

DO use a barrier cream on your hands prior to undertaking dirty jobs – it will protect your skin from infection as well as making the dirt easier to remove afterwards; but make sure your hands aren't left slippery. Note that long-term contact with used engine oil can be a health hazard.

DO keep loose clothing (cuffs, tie etc) and long hair well out of the way of moving mechanical parts.

DO remove rings, wristwatch etc. before working on the vehicle – especially the electrical system.

DO keep your work area tidy – it is only too easy to fall over articles left lying around.

DO exercise caution when compressing springs for removal or installation. Ensure that the tension is applied and released in a controlled manner, using suitable tools which preclude the possibility of the spring escaping violently.

DO ensure that any lifting tackle used has a safe working load rating adequate for the job.

DO get someone to check periodically that all is well, when working alone on the vehicle.

DO carry out work in a logical sequence and check that everything is correctly assembled and tightened afterwards.

DO remember that your vehicle's safety affects that of yourself and others. If in doubt on any point, get specialist advice.

IF, in spite of following these precautions, you are unfortunate enough to injure yourself, seek medical attention as soon as possible.

Asbestos

Certain friction, insulating, sealing, and other products – such as brake linings, clutch linings, gaskets, etc – contain asbestos. *Extreme care must be taken to avoid inhalation of dust from such products since it is hazardous to health.* If in doubt, assume that they *do* contain asbestos.

Fire

Remember at all times that petrol (gasoline) is highly flammable. Never smoke, or have any kind of naked flame around, when working on the vehicle. But the risk does not end there – a spark caused by an electrical short-circuit, by two metal surfaces contacting each other, by careless use of tools, or even by static electricity built up in your body under certain conditions, can ignite petrol vapour, which in a confined space is highly explosive.

Always disconnect the battery earth (ground) terminal before working on any part of the fuel or electrical system, and never risk spilling fuel on to a hot engine or exhaust.

It is recommended that a fire extinguisher of a type suitable for fuel and electrical fires is kept handy in the garage or workplace at all times. Never try to extinguish a fuel or electrical fire with water.

Note: *Any reference to a 'torch' appearing in this manual should always be taken to mean a hand-held battery-operated electric lamp or flashlight. It does **not** mean a welding/gas torch or blowlamp.*

Fumes

Certain fumes are highly toxic and can quickly cause unconsciousness and even death if inhaled to any extent. Petrol (gasoline) vapour comes into this category, as do the vapours from certain solvents such as trichloroethylene. Any draining or pouring of such volatile fluids should be done in a well ventilated area.

When using cleaning fluids and solvents, read the instructions carefully. Never use materials from unmarked containers – they may give off poisonous vapours.

Never run the engine of a motor vehicle in an enclosed space such as a garage. Exhaust fumes contain carbon monoxide which is extremely poisonous; if you need to run the engine, always do so in the open air or at least have the rear of the vehicle outside the workplace.

The battery

Never cause a spark, or allow a naked light, near the vehicle's battery. It will normally be giving off a certain amount of hydrogen gas, which is highly explosive.

Always disconnect the battery earth (ground) terminal before working on the fuel or electrical systems.

If possible, loosen the filler plugs or cover when charging the battery from an external source. Do not charge at an excessive rate or the battery may burst.

Take care when topping up and when carrying the battery. The acid electrolyte, even when diluted, is very corrosive and should not be allowed to contact the eyes or skin.

If you ever need to prepare electrolyte yourself, always add the acid slowly to the water, and never the other way round. Protect against splashes by wearing rubber gloves and goggles.

Mains electricity and electrical equipment

When using an electric power tool, inspection light etc, always ensure that the appliance is correctly connected to its plug and that, where necessary, it is properly earthed (grounded). Do not use such appliances in damp conditions and, again, beware of creating a spark or applying excessive heat in the vicinity of fuel or fuel vapour. Also ensure that the appliances meet the relevant national safety standards.

Ignition HT voltage

A severe electric shock can result from touching certain parts of the ignition system, such as the HT leads, when the engine is running or being cranked, particularly if components are damp or the insulation is defective. Where an electronic ignition system is fitted, the HT voltage is much higher and could prove fat

Routine maintenance

The importance of maintaining a motorcycle conscientiously and carefully cannot be overstressed. Apart from the obvious benefits of safety and economy, a well-maintained motorcycle will generally be ridden with more care and consideration than one that is unkempt or out-of-tune. Therefore it becomes all the more important that necessary tasks be carried out exactly when prescribed and with great care. Not only is a rider risking a great deal more than just money with a badly kept machine, he also runs great risks in safety both to himself and to the machine.

Keeping a machine properly serviced need not be excessively time consuming, although services should be performed regularly and in a professional manner. The various maintenance tasks are described in detail, under their respective mileage and calender headings. It should be remembered that the interval between the various maintenance tasks serves only as a guide. As the machine gets older, is driven hard or is used under particularly adverse conditions, it is advisable to reduce the interval between each check.

If a specific item is mentioned but not described in detail, it will be found in the appropriate Chapter. No special tools are required for the routine maintenance tasks, apart from those found in machine's toolkit. A good set of open ring metric spanners are also very useful, especially for items such as the sump drain plug etc. Their purchase will represent a good investment.

Daily

Check the oil level of the engine-gear unit. This can be done through the sight 'window' at the bottom of the right hand crankcase cover. The correct level is between the two lines marked at the side of the window.

Check the tyre pressures. Always check when the tyres are cool, using an accurate tyre pressure gauge.

Make sure the lights, horn, flashing indicators and speedometer function correctly. The efficient working of these is a statutory requirement by law.

Check that the mirrors are positioned correctly, and that the locknuts are fully tightened.

Make sure the rear chain is correctly adjusted.

Check that the handlebars turn to the right and left smoothly, also that both brakes are working effectively.

Weekly or every 200 miles

In addition to the items already mentioned in the daily check, inspect and adjust the final drive chain.

Check and if necessary adjust both brakes, also the level of the brake fluid in the master cylinder. Top up if necessary.

Check the level of electrolyte in the battery. Examine the wheel spokes for looseness and retension them with a spoke key, if necessary.

Make sure the clutch cable is adjusted correctly and has the necessary free play. Check the oil content of the chain oiler (if fitted).

Monthly or every 1,000 miles

Complete all the previous checks mentioned in the daily and weekly service, then attend to the following items.

Change the engine oil. Note that the oil filter element must be changed every 2,000 miles or, if the engine has been rebuilt, after the first 1,000 miles. This should be accomplished whilst the engine is drained of oil. Wipe off any metal filings that may be attached to the sump drain plug, replace and tighten the plug.

RM1 Filling master cylinder

RM2 Topping up engine oil

Check all the nuts and bolts on the machine, including the cylinder head bolts and the exhaust pipe clamps.

Check and if necessary lubricate the control cables.

Six monthly or every 3,000 miles

Complete all the checks listed in the daily, weekly and 200 mile monthly services and then the following:

Remove the spark plugs, clean and adjust the points gap. If the electrodes are badly eroded or the insulators cracked or badly fouled, the plugs should be renewed.

Check and, if necessary, adjust the contact breaker points. There are two sets of points; when the points have been reset verify the accuracy of the ignition timing.

Check the valve clearances with the engine cold and adjust if necessary.

Check also the tension of the camshaft chain.

Remove and clean the air filter element.

Remove the filter bowl from the petrol tap and clean both the bowl and the filter. Check also that the fuel lines are free from sediment.

Check the clutch cable for adjustment and readjust if required.

Remove and clean the final drive rear chain, then lubricate before replacement if not of the pre-lubricated type. Check also the condition of both gearbox and rear wheel sprockets.

Make sure all four carburettors are clean and adjusted correctly; check that they are synchronized with each other.

Yearly or every 6,000 miles

Complete all the checks under the daily, weekly, monthly, and six monthly headings, then carry out the following additional tasks:

Remove and clean the engine oil filter at the oil pump pick-up. It is retained by three screws.

Renew the air cleaner element.

Remove the rear wheel and check the condition of the brake shoes.

Remove the front wheel, and if the disc pads are down or close to the red danger line, they should be renewed.

Measure the oil level in the front forks and top up if necessary.

Check and adjust the steering head bearings, also check the action of the steering lock. Lubricate it with a few drops of oil.

Check also the tyres for wear; use a tread gauge to measure the depth of tread. Check the tyres for cracks in the sidewalls, and replace if found, in the interests of safety.

RM3 Check fuse if lights won't work

RM4 Engine maintenance data is located under the dualseat

RM5 Checking the points gap

RM6 Use guide lines on rear fork when adjusting the chain

RM7 Remove air filter element for cleaning

RM8 Checking chain oil tank level

RM9 Topping up chain oil tank

RM10 Cleaning filter gauze and bowl

Quick glance
routine maintenance, adjustments and capacities

Refer to Chapter 7 for information relating to the Z900 and Z1000 models

Contact breaker points	0.012 - 0.016 in. (0.3 to 0.4 mm)
Spark plug gaps	0.028 - 0.031 in. (0.7 to 0.8 mm)
Spark plugs	(Normal use) N.G.K. B-8ES) 14 mm (Very high speed) N.G.K. B-9ES) ¾" Reach.
Fuel tank capacity	3.5 Imp. gallons. (4.2 US gallons) (18 litres)
Front fork oil capacity	169 cc per leg
Engine/gearbox oil capacity	4 litres
Oil Tank (rear chain) capacity	7 Imp. pints (1 US gallon) 0.9 litre (Z1 model only)
Tyre pressures (solo)	26 psi (front), 31 psi (rear)
Tyre pressures (pillion)...	26 psi (front), 36 psi (rear)

Lubrication points

1 CONTROL CABLES
2 HYDRAULIC BRAKE RESERVOIR
3 FRONT FORKS
4 SPEEDOMETER AND TACHOMETER CABLES
5 WHEEL BEARINGS
6 SPEEDOMETER DRIVE
7 CONTACT BREAKER CAM
8 KICKSTART PIVOT
9 SWINGING ARM PIVOT
10 BRAKE PEDAL
11 REAR BRAKE OPERATING ARM
12 WHEEL BEARINGS
13 FINAL DRIVE CHAIN
14 CHAIN OILER (Z1 MODEL)
15 ENGINE/GEARBOX OIL

Recommended lubricants

Engine/gearbox unit	SAE 10W/40, 10W/50 or 20W/50 motor oil
Chain oil tank	SAE 90 gear oil
Telescopic forks	SAE 10 fork oil
Hydraulic brake, front master cylinder	DOT 3 or DOT 4
Control cables	Light oil
Grease nipples	General purpose grease

Chapter 1 Engine, clutch and gearbox

Contents

General description 1
Operations with engine/gearbox unit in the frame ... 2
Operations with engine/gearbox removed 3
Method of engine gearbox removal 4
Removing the engine/gearbox unit 5
Dismantling the engine and gearbox: general 6
Dismantling the engine and gearbox: removing the camshafts, cylinder head, and cylinder block 7
Dismantling the engine/gearbox: removing the pistons and piston rings 8
Dismantling the engine and gearbox: removing the contact breaker assembly 9
Dismantling the engine and gearbox: removing the clutch ... 10
Dismantling the engine and gearbox: removing the alternator and starter motor 11
Dismantling the engine and gearbox: removing the final drive sprocket 12
Dismantling the engine and gearbox: removing the oil filter 13
Dismantling the engine and gearbox: separating the crankcases 14
Removing the crankshaft assembly and gear clusters ... 15
Dismantling the gearbox: dismantling the gear clusters ... 16
Examination and renovation general 17
Crankshaft, big ends, main bearings: examination and renovation 18
Oil seals: examination and replacement 19
Cylinder block: examination and renovation 20
Pistons and piston rings: examination and renovation ... 21
Cylinder head: examination and renovation 22

Valves, valve seats and valve guides: examination and renovation 23
Camshafts, tappets, and camshaft sprockets: examination and renovation 24
Clutch and rear chain oil pump: examination and renovation 25
Gearbox components: examination 26
Reassembling the cylinder head 27
Engine and gearbox reassembly: general 28
Engine and gearbox reassembly: replacing the crankshaft ... 29
Engine and gearbox reassembly: rebuilding the gearbox ... 30
Engine and gearbox: rebuilding the clutch, alternator, and contact breaker 31
Engine and gearbox reassembly: replacing the pistons and cylinder block 32
Engine and gearbox reassembly: replacing the cylinder head and camshafts, retiming the engine 33
Engine and gearbox reassembly: adjusting the valve clearances 34
Engine and gearbox reassembly: replacing the carburettors 35
Engine and gearbox reassembly: replacing the engine and gearbox into the frame 36
Starting and running the rebuilt engine unit 37
Taking the rebuilt machine on the road 38
Fault diagnosis: engine 39
Fault diagnosis: clutch 40
Fault diagnosis: gearbox 41

Specifications

Engine

Type	4 cylinder transverse D.O.H.C. in-line 4 stroke air cooled
Bore and stroke	66 x 66 mm
Displacement	903 cc.
Compression ratio	8.5 to 1.
Maximum horsepower	82 h.p. @ 8,500 r.p.m.
Maximum torque	54.3 ft lb @ 7,000 r.p.m. (7.5 kg m @ 7,000 r.p.m.)
Cylinder block	Aluminium alloy, steel liners.
Cylinder head	Aluminium alloy, angled plug threads.

Pistons

Type	Aluminium alloy
Oversizes available	0.5 mm (0.020 inch) and 1.0 mm (0.040 inch)

Piston rings:

Number per piston	Two compression and one oil control
End gap	0.008 — 0.016 in. (0.2 — 0.4 mm) wear limit 0.028 in.

Groove clearance:

Top ring ...	.	...	...	...	...	...	...	0.0018 — 0.0031 in. (0.045 — 0.080 mm)
Second ring ...	...	...	...	...	...	...	...	0.0004 — 0.0020 in. (0.010 — 0.050 mm)
Oil ring ...	...	...	...	...	...	...	...	Same

Cylinder block

Bore wear limit ...	...	...	...	1 mm (0.03937 inches) If cylinder rebore exceeds 1 mm cylinder block will have to be replaced with a new one.

Valves:

Valve stem diameter: Inlet ...	...	...	...	...	...	6.965 — 6.880 mm. (0.2742 — 0.2748 in.)
Valve stem diameter: exhaust	...	...	...	...	...	6.955 — 6.970 mm. (0.2738 — 0.2744 in.)
Wear limit inlet:	...	...	...	...	...	6.86 mm (0.270 in.)
Wear limit exhaust: ...	...	...	...	...	...	6.85 mm (0.270 in.)

Valve springs:

Free length inner ...	...	...	...	...	...	...	36.0 mm (1.42 inch)
Free length outer ...	...	...	...	...	...	...	39.3 mm (1.55 inch)
Wear limit inner ...	...	...	...	...	...	...	35.0 mm (1.38 inch)
Wear limit outer ...	...	...	...	...	...	...	38.0 mm (1.50 inch)
Valve clearance engine cold ...	...	...	...	...	...	0.002 — 0.004 inch (0.05 — 1.0 mm)
Valve timing: Inlet opens ...	...	...	...	...	...	30° BTDC
Inlet closes ...	...	...	...	...	...	70° ABDC
Exhaust opens ...	...	...	...	...	...	70° BBDC
Exhaust closes ...	...	...	...	...	...	30° ATDC

Clutch

Number of inserted plates ...	...	...	...	...	...	8
Number of plain plates ...	...	...	...	...	...	7
Thickness of the inserted plates ...	...	...	...	...	...	3.7 mm — 3.9 mm (0.146 — 0.154 in.)
Wear limit of inserted plates ...	...	...	...	...	...	3.4 mm (0.134 in.)
Clutch springs - free length ...	...	...	...	...	...	33.8 mm (1.33 in.)
Clutch springs - wear limit ...	...	...	...	...	...	32.3 mm (1.27 in.)

Torque wrench settings

							lb ft	kg m
Cylinder head nuts ...	...	...	...	...	...	...	25	3.5
Exhaust manifold nuts ...	...	...	...	...	...	...	100 ins - lbs	1.2
Cylinder head bolts ...	...	...	...	...	...	...	105 ins - lbs	1.2
Cylinder head cover bolts ...	...	...	...	...	...	102 ins - lbs	1.2
Crankshaft cap bolts ...	...	...	...	...	...	...	18	2.5
Crankcase nuts (6 o mm) ...	...	...	...	...	...	70 ins - lbs	0.8
Crankcase nuts (8 o mm) ...	...	...	...	...	...	18	2.5
Camshaft cap bolts ...	...	...	...	...	...	...	104 ins - lbs	1.2
Clutch centre hub nut ...	...	...	...	...	...	87 — 108	12 — 15
Generator rotor to crankshaft ...	...	...	...	...	...	18	2.5
Engine oil pump bolts ...	...	...	...	...	...	70 ins - lbs	0.8
Engine sprocket nut ...	...	...	...	...	...	...	87 - 108	12 - 15

1 General description

The engine unit fitted to the Kawasaki Z1 series is of the 4 cylinder in-line type, fitted transversely across the frame. The valves are operated by double overhead camshafts driven off the crankshaft by a centre chain. The two camshafts are located in the cylinder head casting, and the camshaft chain drive operates through a cast-in tunnel between the four cylinders. Adjustment of the chain is effected by a chain tensioner, fitted to the rear of the cylinder block.

The engine/gear unit is of aluminium alloy construction, with the crankcase divided horizontally.

The Z1 series have a wet sump, pressure fed lubrication system, which incorporates a gear driven oil pump, an oil filter, a safety by-pass valve, and an oil pressure switch.

Oil vapours created in the crankcase are vented through an oil breather to the air cleaner hose where they are recirculated into the crankcase, providing an oil tight system.

The oil pump is a twin shaft dual rotor unit, which is driven off the crankshaft by a gear.

An oil strainer is fitted to the intake side of the oil pump, which serves to protect the pump mechanism from any impurities in the oil that might cause damage.

The oil filter unit which is housed in the sump is an alloy canister with a paper element. As the oil filter becomes clogged with impurities, its ability to operate efficiently is reduced, and when it becomes so clogged that it begins to impede the oil flow, the by-pass valve opens and routes the oil around the filter. This of course results in unfiltered oil being circulated throughout the engine, a condition that will be avoided if the filter element is changed at the prescribed intervals.

The lubrication flow is as follows: Oil is drawn from the sump through the oil strainer to the pump, then it passes through the oil filter (or around it if the by-pass valve is in operation) to the pipe in which the oil pressure switch is mounted. It is then routed through three branch systems. The first system lubricates the crankshaft main bearings and crankpins. The oil is thrown by the crankshaft's rotating motion onto the cylinder walls providing the splash lubrication for the pistons. The oil then drips down into the sump, to be recirculated.

The second system lubricates the cylinder head assembly. Oil flows up through passages in the cylinder block, through the camshaft bushes, down over the cams, through the cam lifters or (tappets) and back to the sump by way of holes in the base of the tappets, and the cam chain tunnel in the cylinder head.

The third system feeds the transmission bearings and then drains back to the sump for recirculation.

The engine is built in-unit with the gearbox. This means that when the engine is completely dismantled, the clutch and gearbox are dismantled too. This task is made easy by arranging the crankcase to separate horizontally.

2 Operations with the engine/unit in the frame

1 It is not necessary to remove the engine from the frame to carry out certain operations; in fact it can be an advantage. Tasks that can be carried out with the engine in situ are as follows;

  a) *Removal and replacement of the clutch.*
  b) *Removal and replacement of the flywheel generator.*
  c) *Removal and replacement of the generator rotor.*
  d) *Removal and replacement of the carburettors.*
  e) *Removal and replacement of the starter motor.*
  f) *Removal and replacement of the rear chain oil pump.*

2 When several tasks have to be undertaken simultaneously, it will probably be advantageous to remove the complete engine unit from the frame, an operation that should take about an hour and a half. This gives the advantage of much better access and more working space.

3 Operations with the engine/gearbox unit removed from the frame

  a) *Removal and replacement of the cylinder head unit.*
  b) *Removal and replacement of the cylinder block.*
  c) *Removal and replacement of the pistons.*
  d) *Removal and replacement of the crankshaft assembly.*
  e) *Removal and replacement of the main bearings.*
  f) *Removal and replacement of the gear clusters and selectors.*
  g) *Removal and replacement of the kickstart mechanism, gearbox bearings, and gear change mechanism.*

4 Method of engine/gearbox removal

As mentioned previously, the engine and gearbox are built in-unit and it is necessary to remove the complete engine unit to gain access to either component.

The engine unit is secured to the frame with three long bolts and twelve short mounting bolts. After these have been removed, and the necessary electrical connections disconnected, together with the carburettor fuel pipes, plug leads and exhaust system, the engine is ready for removal. Dismantling of the engine unit can only be accomplished after the engine unit has been removed from the frame and refitting cannot take place until the engine unit has been reassembled.

5 Removing the engine/gearbox unit

1 Place the machine firmly on its centre stand so that it stands on a smooth, level surface. The ideal position for working is to place the machine on a stout wooden stand about 18 inches high, resting on its centre stand.

2 Make sure you have a clean, well-lit place to work in and a good set of tools. You will need at least three sizes of crosshead (Phillips) screwdrivers, small, medium and large, and plenty of clean lint free rag.

3 Remove the oil sump plug, also the oil filter drain plug, and drain the oil into a suitable tray. Approximately one gallon of oil will drain off.

4 Disconnect and remove the battery. The battery is located beneath the dualseat, in a cradle compartment. It should be lifted straight up, being careful not to spill the contents.

5 Drain and remove the petrol tank. The tank is held by a rubber clip at the rear that engages in a lip welded onto the rear of the tank. Unhook the rubber band and pull the fuel tank off, toward the rear.

6 Remove the finned clips from all four exhaust pipes. They are held by two nuts per clamp, secured to studs fitted into the cylinder head. It is a good idea to soak these nuts in penetrating oil before undoing them, to safeguard against breakage of the studs in the cylinder head. Remove the mounting bolts at the rear of the silencers and push the silencers forward, removing them as a pair from each side. Remove the right hand rider's footrest by removing the two retaining bolts.

7 Detach the rear brake cable from its anchor point near to the rear of the brake pedal.

5.4 Disconnect and remove the battery

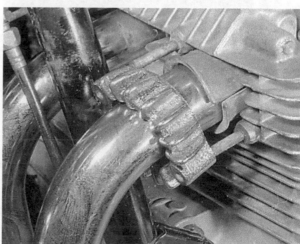

5.6 Remove the finned clips

5.6A Bolt secures silencers

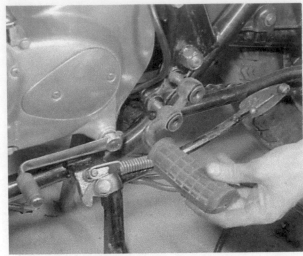

5.6B Remove footrests

5.7 Detach rear brake rod

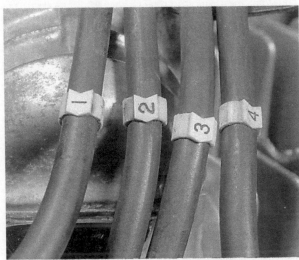

5.8 Plug leads are numbered

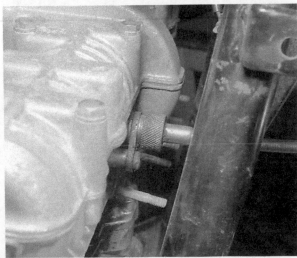

5.8A Unscrew tachometer cable

5.9 Remove battery earth lead

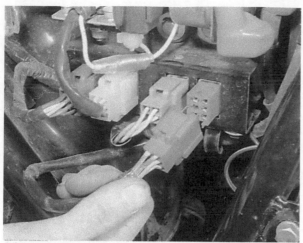
5.9A Unplug the blue connector

8 Take off the spark plug leads, and tuck them out of the way. Remove the tachometer cable from directly in front of the cylinder head by undoing the knurled nut.
9 Remove the negative battery lead from its earth on the engine, unplug the blue connector (note you have to squeeze this connector as the sides are spring loaded, this is a safeguard to hold them tight when assembled).
10 Disconnect the black wire and the green wire that join the ignition coils to the contact breaker points. Remove the starter wire from the relay terminal in front of the electrical panel and near to the rear right hand frame tube. Remove the air cleaner assembly by undoing the crossheaded screws on the carburettor hoses.
11 Loosen all eight clamps on the carburettors and pull the carburettor assembly off to the rear. Loosen the adjusters on the two throttle cables and unhook the throttle cables from the pulley.
12 Remove the rubber hose from the oil breather assembly situated at the rear of the engine, and remove the oil breather assembly. Take off the left hand rider's footrest, and remove the gear change pedal complete.

5.10 Disconnect coil wires

5.10A Remove starter cable at solenoid

5.10B Remove air cleaner assembly

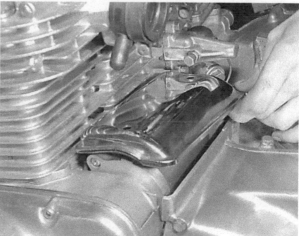
5.13 Take off starter motor cover

5.13A Remove chain oil pump cover

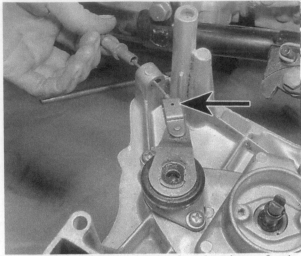

5.13B Remove split pin from hole in release lever (arrowed) and disconnect clutch cable

5.13C Unhook spring from stopswitch

5.13D Remove gearbox sprocket cover

5.13E Taking off gearbox sprocket

5.15 Remove the lower brackets

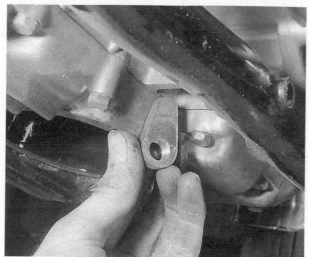
5.15A Threaded plates fit up into crankcase

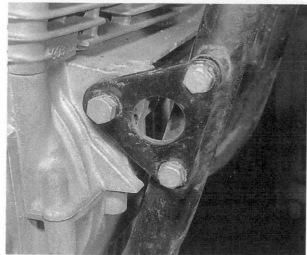
5.16 Remove front engine bracket

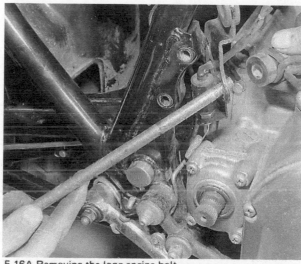
5.16A Removing the long engine bolt

5.16B Engine mounting at rear

5.16C Note spacer at top rear mounting

5.16D Engine unit ready for removal

5.17 Raise engine up...

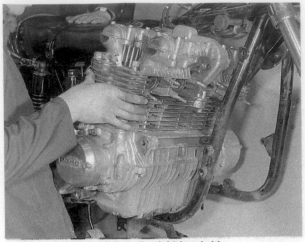

5.17A ...and remove it from the right-hand side

13 Take off the starter motor and gasket. Remove the rear chain oil pump cover (if fitted), pull the hose off the pump and plug the hose with a screw to prevent oil leakage. Remove the chain cover, pull the split pin out from the clutch release lever and unhook the clutch cable. Unhook the spring from the stopswitch and take off the switch. Remove the gearbox sprocket cover and remove the gearbox sprocket nut after preventing rotation by applying the rear brake. Disengage the sprocket from the rear chain and allow the latter to hang in position. It need not be removed completely.

14 Loosen the locknut on the rear brake pedal position adjusting bolt and turn the bolt downwards until it holds the brake pedal well out of the way.

15 Use a small jack or alternatively a block to put under the crankcase to take the weight off the engine bolts. Remove the nuts off the three long engine bolts and completely remove the short engine mounting bolts from the lower centre mounting on each side of the machine. These are bolted to threaded plates that fit up into the bottom of the crankcase.

16 Remove the right hand rear engine mounting bolts, remove the centre and front right hand side engine mounting brackets, then remove the three long engine mounting bolts one at the front and two at the rear of the engine.

17 Level the engine and slowly lift it straight up about 1 inch (25 mm), then move it to the right slightly so that the rear of the engine slips over the rear lower mounting bracket. Raise the front of the engine a little at this stage, so that it will clear the frame, then lower the left hand side of the engine and pull the engine out of the frame cradle, lifting diagonally upwards towards the right hand side. It should be noted that two persons are needed to handle an engine of this size.

6 Dismantling the engine/gearbox: general

1 Before commencing work on the engine unit, the external surfaces should be cleaned thoroughly. A motorcycle engine has very little protection from the hazards of road grit and other foreign matter, due to it having to be constructed to take advantage of air cooling.

2 There are a number of proprietary cleaning solvents on the market including Jizer and Gunk. It is best to soak the parts in one of these solvents, using a cheap paint brush. Allow the solvent to penetrate the dirt, and afterwards wash down with water, making sure not to let water penetrate the electrical system or get into the engine, as many parts are now more exposed.

3 Have a good set of tools ready, including a set of open ring metric spanners (these have a ring one end and are open ended at the other end), a few metric socket spanners of the smaller sizes,

and an impact screwdriver with a selection of bits for the crosshead screws. If one is not available, a crosshead screwdriver with a T handle fitted can sometimes be used as a substitute. Work on a clean surface and have a supply of clean lint free rag available.

4 Never use force to remove any stubborn part unless specific mention is made of this requirement in the text. There is invariably good reason why a part is difficult to remove, often because the dismantling procedure has been taken out of sequence.

7 Dismantling the engine and gearbox: removing the camshafts, the cylinder head, and cylinder block

1 Place the engine on the bench and remove the camshaft cover. To remove the camshafts, slacken the camshaft chain tensioner and undo the bolts holding the camshaft cover. Remove the chain guide sprocket located at the top of the engine, unbolt the camshaft caps, and mark the split bushes under each cap to show their location. This is important as any bushes that have to be re-used must go back in their original locations.

2 The camshafts can now be removed by slipping them under the camshaft chain. Unscrew the tachometer cable and take out the tachometer pinion, so that it does not cause damage to the camshaft worm during camshaft installation at a later stage.

3 The cylinder head is now ready for removal. Start by removing the two bolts at each end of the cylinder head, then undo the twelve cylinder head nuts with a suitable socket wrench.

4 Remove all the valve tappets and shims. Keep them separate for installation in their original locations. This is most important. Remove the cylinder head with a soft headed mallet to break it free from its seat, if necessary.

5 The cylinder block can now be removed by tapping upwards with a soft headed mallet. As the cylinder block is raised it is best to pad the crankcase mouths with clean rag to eliminate any foreign particles from falling into the crankcase, especially if a bottom end overhaul is not contemplated. Catch the pistons as they emerge from the cylinder bores, and lower the camshaft chain through the centre tunnel on a piece of wire.

8 Dismantling the engine and gearbox: removing the pistons and piston rings

1 Remove the circlips from the pistons by inserting a screwdriver (or a piece of welding rod chamfered one end), through the groove at the rear of the piston. Discard them. Never

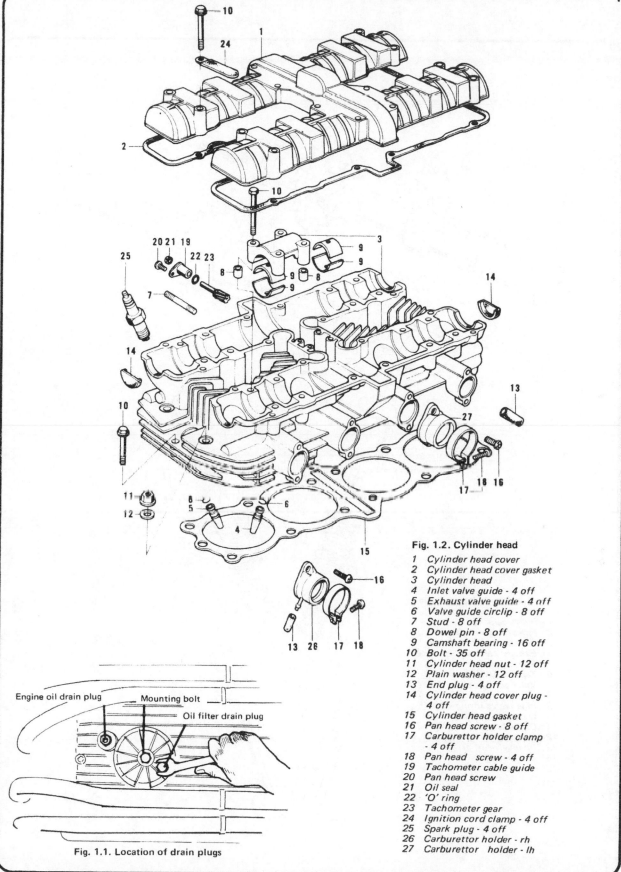

Fig. 1.2. Cylinder head

1 Cylinder head cover
2 Cylinder head cover gasket
3 Cylinder head
4 Inlet valve guide - 4 off
5 Exhaust valve guide - 4 off
6 Valve guide circlip - 8 off
7 Stud - 8 off
8 Dowel pin - 8 off
9 Camshaft bearing - 16 off
10 Bolt - 35 off
11 Cylinder head nut - 12 off
12 Plain washer - 12 off
13 End plug - 4 off
14 Cylinder head cover plug - 4 off
15 Cylinder head gasket
16 Pan head screw - 8 off
17 Carburettor holder clamp - 4 off
18 Pan head screw - 4 off
19 Tachometer cable guide
20 Pan head screw
21 Oil seal
22 'O' ring
23 Tachometer gear
24 Ignition cord clamp - 4 off
25 Spark plug - 4 off
26 Carburettor holder - rh
27 Carburettor holder - lh

Engine oil drain plug   Mounting bolt

Oil filter drain plug

Fig. 1.1. Location of drain plugs

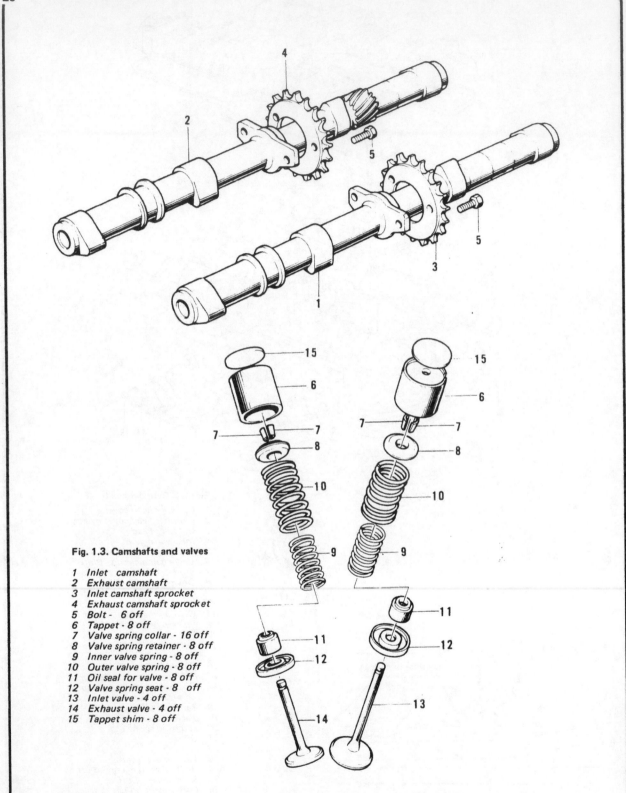

20

Fig. 1.3. Camshafts and valves

1   Inlet   camshaft
2   Exhaust camshaft
3   Inlet camshaft sprocket
4   Exhaust camshaft sprocket
5   Bolt - 6 off
6   Tappet - 8 off
7   Valve spring collar - 16 off
8   Valve spring retainer - 8 off
9   Inner valve spring - 8 off
10 Outer valve spring - 8 off
11 Oil seal for valve - 8 off
12 Valve spring seat - 8 off
13 Inlet valve - 4 off
14 Exhaust valve - 4 off
15 Tappet shim - 8 off

7.1 Remove the camshaft cover bolts

7.1A and lift off the cover

7.1B Remove chain guide sprocket

7.1C Unbolt the camshaft caps

7.1D Take off the caps and bushes

7.2 Removing camshaft, exposes tappets

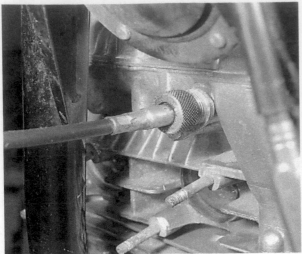
7.2A Unscrew tachometer cable

7.3 Small bolts at each end of cylinder head

7.4 Lift cylinder head off studs

7.5 Removing cylinder block

8.1 Prise in groove to release circlip

8.2 Tap gudgeon pin out lightly

9.1 Remove contact breaker cover

9.1A Lift off assembly complete

9.2 Taking off the backplate

10.1 Remove the cover and gasket

10.2 Remove the pressure plate bolts

10.3 Lift out the clutch plates

10.3A One way of locking the hub

10.4 Pull off the hub

10.5 Take off the kickstart cover

11.1 Take off cover with coils attached

11.1A Remove the bolt and washer

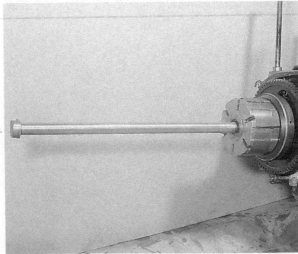

11.1B Rear wheel spindle will remove rotor

re-use old circlips during the rebuild.

2 Using a drift of suitable diameter, tap each gudgeon pin out of position by supporting each piston and connecting rod in turn. Mark each piston inside the skirt so that it is replaced in the same position. If the gudgeon pins are a tight fit into the connecting rods, it is advisable to warm the pistons. One way is to soak a rag in very hot water, wring the water out and wrap the rag round the piston very quickly. The resultant expansion should ease the grip of the piston bosses.

9 Dismantling the engine and gearbox: removing the contact breaker assembly

1 Remove the contact breaker cover by undoing the two screws that hold the cover on the front right hand side of the engine. This cover has D.O.H.C. stamped on it. Then remove the three screws holding the contact breaker assembly with condensers attached, lift off the plate, separating the wires by pulling them out of the connector.

2 Remove the bolt that holds the auto-advance mechanism to the shaft, and remove the advance mechanism with the cam. Now undo the six screws holding the backplate. Note there are two dowels that locate into the side of the crankcase. After this plate is removed, the crankshaft is exposed.

10 Dismantling the engine and gearbox: removing the clutch

1 After making sure that all the oil has been drained from the clutch housing, as previously mentioned at the beginning of this Chapter, undo the eight crosshead screws that retain the clutch cover, and remove both cover and gasket.

2 Remove the five pressure plate bolts, the clutch springs, and the pressure plate. Take out the clutch mushroom pushrod, and tilt the engine over, so that the steel ball will fall out (otherwise known as the release bearing).

3 Remove the clutch plates (there are eight friction, and seven plain plates). The clutch plates can be removed by using a hooked piece of wire to lift them out. Then remove the clutch hub centre nut by either using the special tool that holds the hub still, or a suitable substitute.

4 Remove the outer washer, the hub, and the inner washer. The clutch outer case cannot be removed until the crankcases are separated.

5 The kickstart cover can now be removed from the rear of the clutch case, by undoing the four crosshead screws. The kickstart spring can be pulled out with a pair of long nosed pliers.

11 Dismantling the engine and gearbox: removing the alternator and starter motor

1 Turn to the opposite side of the engine (left hand), remove the six crosshead screws that hold on the outer cover. The maker's name ('KAWASAKI') is stamped on the cover. After this cover is removed together with its gasket, the rotor will be exposed. If a rotor extractor is not available, the rear wheel spindle can be screwed into the thread of the rotor, after the rotor securing bolt and washer have been removed. By tapping outwards under the head of the spindle after it has been tightened, the rotor should come off, with the starter pinion attached. **Note: On no account hammer the rotor.**

2 Take out the two long bolts that fit vertically in the end of the starter motor bracket, unscrew the starter cable from the terminal, and release the cable from the clamp. The starter motor can now be pulled out from the crankcase.

3 The starter pinion with gear can now also be removed.

11.1C Take off rotor with gear

11.1D Do not lose Woodruff key

11.2 Removing the starter motor

11.3 Take off the starter pinion, and gear

12.1 Disconnect the neutral switch wire

12.2 Note locating pin in shaft

12.3 Take off transmission cover

12.4 Remove gear change mechanism

12.4A Undo bolt and remove detent lever

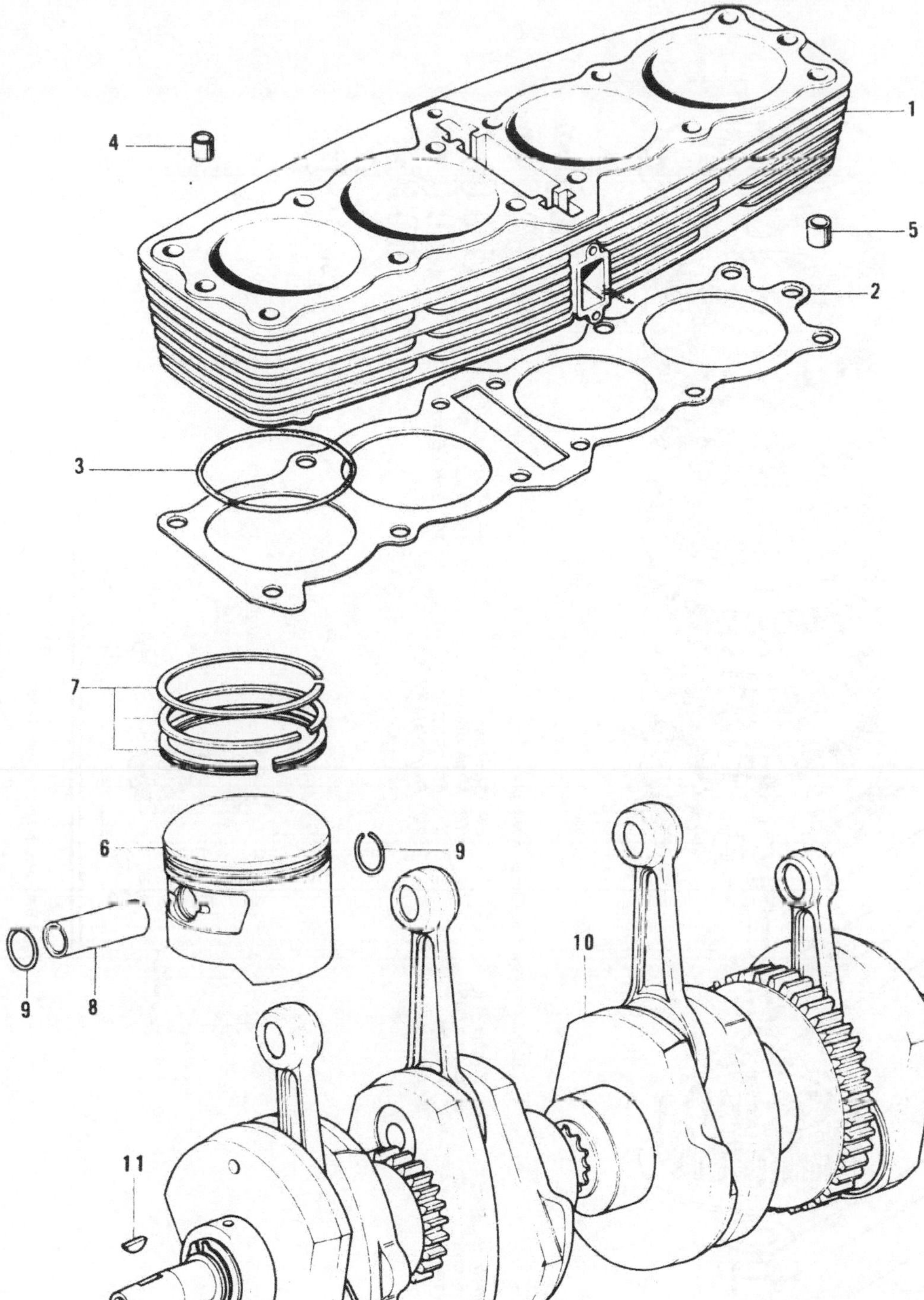

Fig. 1.4. Cylinder, piston and crankshaft

1	Cylinder	6	Piston - 4 off
2	Cylinder base gasket	7	Piston ring set - 4 off
3	'O' ring - 4 off	8	Gudgeon pin - 4 off
4	Dowel pin - 2 off	9	Circlip - 8 off
5	Dowel pin - 2 off	10	Crankshaft assembly
		11	Alternator Woodruff key

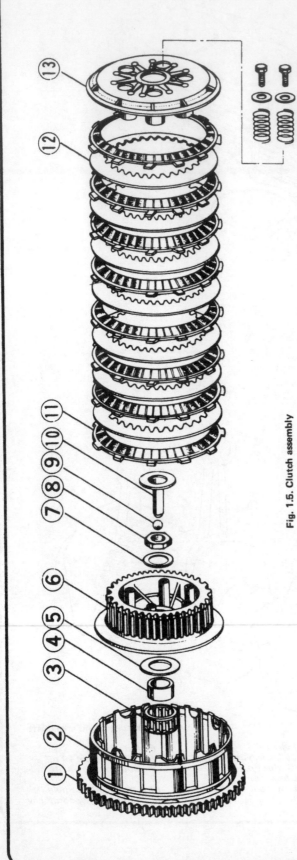

Fig. 1.5. Clutch assembly

1 Reduction drive gear
2 Clutch housing
3 Needle bearing
4 Bush
5 Washer
6 Clutch hub
7 Washer
8 Nut
9 Steel ball
10 Mushroom pushrod
11 Friction plate - 8 off
12 Plain plate - 7 off
14 Clutch spring - 5 off
15 Washer - 5 off
16 Bolt - 5 off

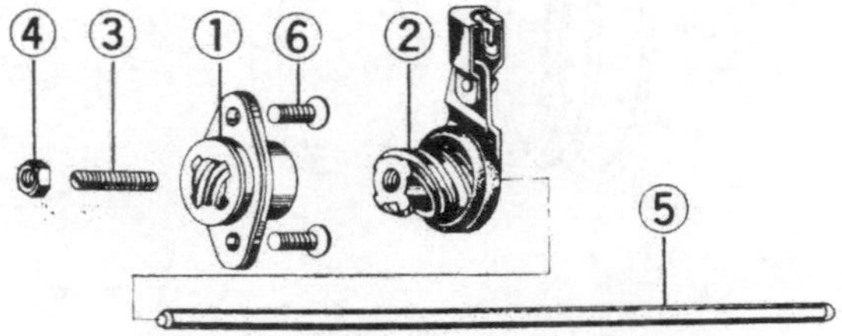

Fig. 1.5A Clutch release mechanism

1 Outer release gear
2 Inner release gear
3 Adjusting screw
4 Lock nut
5 Push rod
6 Screw

12 Dismantling the engine and gearbox: removing the final drive sprocket

1   Pull out the clutch pushrod, disconnect the neutral switch situated at the bottom of the sprocket guard, unscrew the three crosshead screws and remove the sprocket guard.
2   Flatten the bent up ear of the tab washer locking the sprocket nut, undo the sprocket nut while stopping the sprocket from turning (one way of doing this is to put a piece of chain onto the sprocket and then to hold both ends of the chain in a vice). Remove the sprocket.
3   Take off the transmission cover and gasket, remove the sprocket distance collar and take off the 'O' ring that fits behind the collar. Note: The 'O' ring is to stop oil leakage from the crankcase.
4   Remove the gear change detent lever mounting bolt, unhook the lever spring, and remove the lever.

13 Dismantling the engine and gearbox: removing the oil filter

1   The oil filter is located underneath the sump. The filter has a drain plug, that should normally be removed when changing the engine oil. However, as we are taking the filter out complete it can be left in and the whole filter plate removed by undoing the centre mounting bolt. The filter assembly with the element will then come down from the crankcase. Take off the rubber gaskets and remove the filter element.

14 Dismantling the engine and gearbox: separating the crankcases

1   Take out the crankcase bolts from the top side of the engine (two of them at the centre position just behind the crankshaft have cable clips attached). Remove the centre camchain guide roller.
2   Turn the engine upside down on the bench, and remove the oil strainer, oil pan, and gasket. Undo the three bolts that retain the oil pump. The pump can now be removed. Remove the seventeen 6 mm and the eight 8 mm bolts from the crankcase bottom half. Do not remove the four bolts that hold the crankshaft.
3   There are three jack bolt positions in the lower crankcase. To separate the crankcases, screw three 8 mm bolts into these holes (two at the front and one at the rear position of the engine), and the crankcases will separate. Take care to screw in the bolts evenly, half a turn at a time, to make sure the crankcase separates evenly.

15 Dismantling the engine and gearbox: removing the crankshaft assembly and gear clusters

1   To remove the crankshaft assembly undo the four bolts that hold the centre crankshaft bearing cap. This bearing cap is bored 'in line' with the crankcase. There is only one position for it, with the arrow pointing towards the front of the engine.
2   Take out the crankshaft assembly from the upper crankcase half and slip off the cam-chain.
3   The gearbox mainshaft complete with gear cluster, and the clutch housing can now be removed.
4   Lift out the layshaft with gear cluster, also the kickstart shaft and pinion.
5   Lightly tap out the selector fork rod and remove the two selector forks. Remove the detent arm, bend back the locking tab of the lockwasher, and remove the gear change fork pin.
6   Bend back the locktab on the selector drum bolt, then unscrew and remove the selector drum positioning bolt. The cap bolt need not be removed from the top of the positioning bolt.
7   Pull the selector drum out of the crankcase along with the third gear selector fork.

13.1 Take off cover with bolt

14.1 The camchain roller

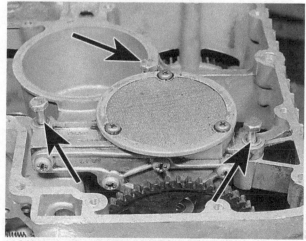

14.2 Undo the three bolts to remove the oil pump

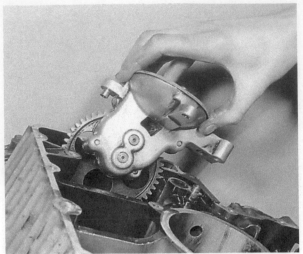

14.2A Withdraw the oil pump

14.3 Separating the crankcases

14.3A The bottom crankcase removed

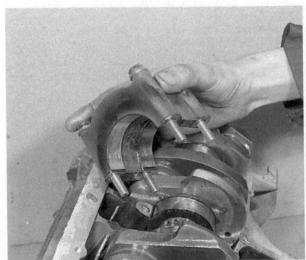

15.1 Undo the four bolts, to remove cap

15.2 Lift up crankshaft assembly

15.3 Take out gear cluster and clutch

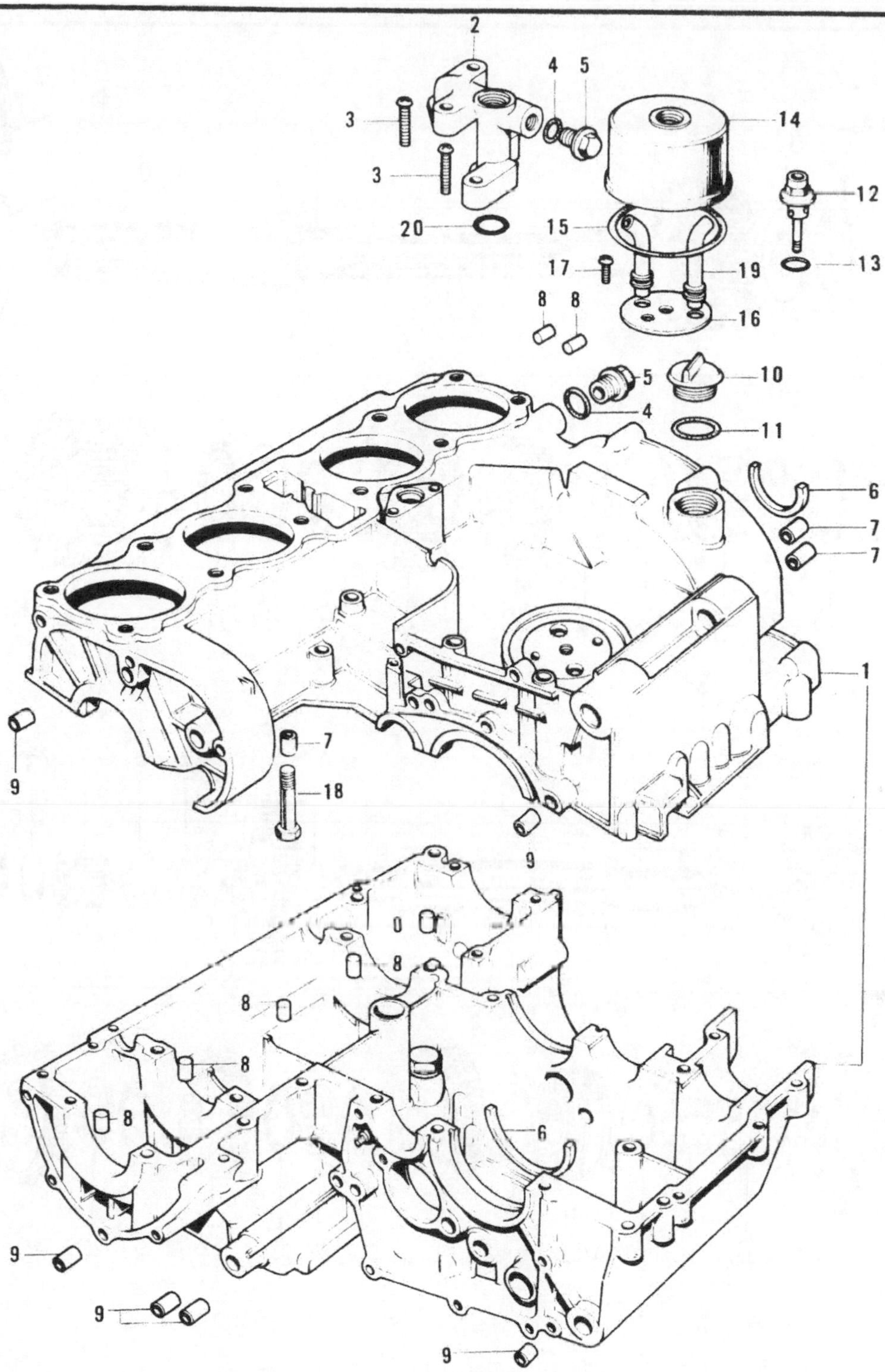

Fig. 1.6. Crankcase assembly

1 Crankcase complete	6 Bearing set ring - 2 off	11 'O' ring	16 Breather plate
2 Oil pressure switch body	7 Dowel pin - 4 off	12 Bolt	17 Pan head screw - 2 off
3 Pan head screw - 4 off	8 Dowel pin - 8 off	13 'O' ring	18 Bolt - 4 off
4 'O' ring - 2 off	9 Dowel pin - 6 off	14 Breather body	19 Tube - 2 off
5 Plug - 2 off	10 Oil filler plug	15 'O' ring for breather body	20 'O' ring - 2 off

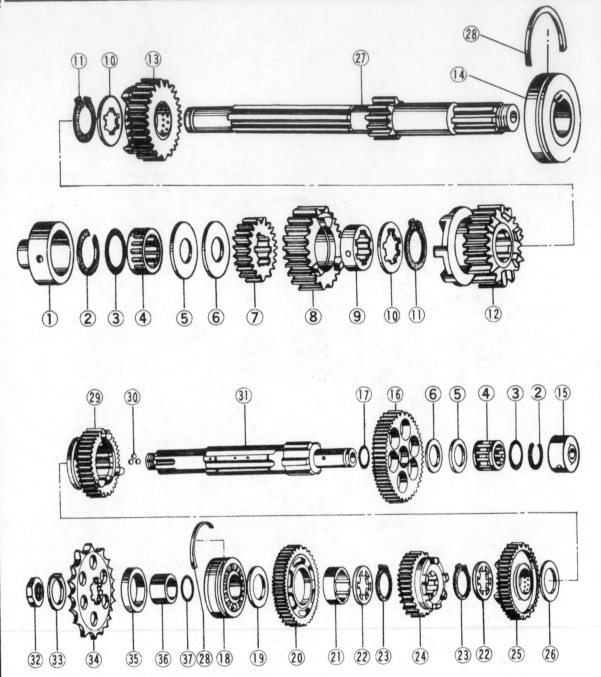

Fig. 1.7 Gearbox components

1 Bush	11 Retaining ring	21 Bush 2nd gear	31 Layshaft
2 Retaining ring	12 Mainshaft 3rd gear	22 Splined washer	32 Nut
3 Shim	13 Mainshaft 4th gear	23 Retaining ring	33 Lock washer
4 Needle roller bearing	14 Bearing	24 Layshaft 5th gear	34 Engine sprocket
5 Washer	15 Bush	25 Layshaft 3rd gear	35 Oil seal
6 Washer	16 Layshaft 1st gear	26 Washer	36 Engine sprocket collar
7 Mainshaft 2nd gear	17 Shim	27 Mainshaft	37 'O' ring
8 Mainshaft 5th gear	18 Bearing	28 Set ring	
9 Bush 5th gear	19 Washer	29 Layshaft 4th gear	
10 Washer	20 Layshaft 2nd gear	30 Steel balls	

15.4 Lift out layshaft cluster

15.4A Remove kickstart shaft assembly

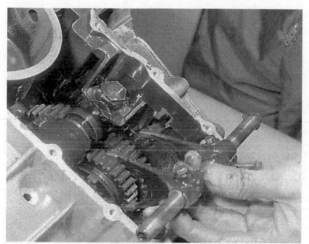

15.5 Remove selector forks

15.6 Unscrew selector fork bolt

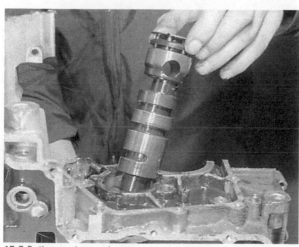

15.7 Pull out selector drum

16 Dismantling the gearbox: dismantling the gear clusters

1   To dismantle the mainshaft, first remove the outer bush, then the retaining ring and shim and next the needle roller bearing and the two spacers. After this take off the 2nd gear pinion and the 5th gear pinion with the 5th gear bush, the splined washer and external circlip. Next remove the sliding 3rd gear, the circlip and splined washer, and the 4th gear with bush. On the opposite end of the shaft, next to the fixed pinion, is a roller bearing. This can be removed if it has to be renewed, using a bearing puller. Take off the split set ring halves.

2   Start to dismantle the layshaft gears by first removing the bush, retaining ring and shim, and then the needle roller bearing and two spacing washers. Next remove the large 1st gear and the shim located behind the gear pinion.

3   Rotate the layshaft while removing the 4th gear, so that the three balls will move. Then pull the bearing from the shaft with a bearing puller, not forgetting to remove the bearing set rings halves first. Remove the washer and the large 3rd gear pinion, then remove the 2nd gear bush, the splined washer and the external circlip.

4   Take off the sliding 5th gear pinion, the external circlip, and the second splined washer. Remove the large 3rd gear with the bush and lastly, remove the washer.

16.1 The gear clusters, before dismantling

16.1A First remove outer bush

16.1B Use circlip pliers on retaining ring

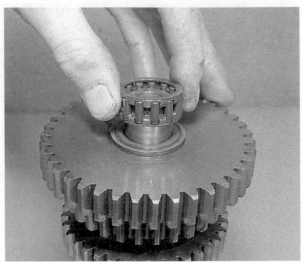

16.1C Then remove needle roller

16.1D Take off ball bearing, on other end

16.1E and then spacer

16.1F Remove snap ring

16.1G Take off sliding gear

16.1H Remove bush with dowel hole

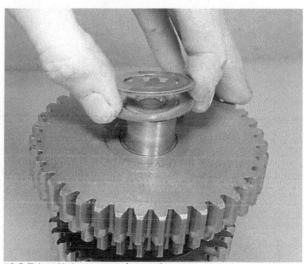

16.2 Take off the two spacing washers

16.3 Careful not to misplace the three balls in fourth gear

16.3A Pull off ball bearing

16.3B Take off third gear pinion

16.3C Remove bush

16.3D Then splined washer and circlip

16.4 Take off sliding gear

17 Examination and renovation: general

1   Before examining the parts of the dismantled engine unit for wear, it is essential that they should be cleaned thoroughly. Use a petrol/paraffin mix to remove all traces of old oil and sludge which may have accumulated within the engine.

2   Examine the crankcase castings for cracks or other signs of damage. If a crack is discovered it will require a specialist repair.

3   Examine carefully each part to determine the extent of wear, checking with the tolerance figures listed in the Specifications section of this Chapter. If there is any question of doubt play safe and renew.

4   Use a clean lint free rag for cleaning and drying the various components. This will obviate the risk of small particles obstructing the internal oilways, and causing the lubrication system to fail.

18 Crankshaft big ends and main bearings: examination and renovation

1   Failure of the big end bearings is invariably accompanied by a knock within the crankcase which progressively becomes worse.

Some vibration will also be experienced.

2   There should be no vertical play whatsoever in the big end bearings, after the oil has been washed out. If even a small amount of vertical play is evident, the bearings are due for replacement. (A small amount of endfloat is both necessary and acceptable). Do not continue to run the machine with worn big end bearings, for there is a risk of breaking the connecting rods or crankshaft.

3   The construction of the crankshaft assembly precludes the possibility of a repair by the private owner or even most agents. In the majority of cases it will be necessary to obtain a replacement assembly complete, from a KAWASAKI agent.

4   Failure of the main bearings is usually evident in the form of an audible rumble from the crankcase, accompanied by vibration, which is felt through the footrests and handlebars in extreme cases.

5   The crankshaft main bearings are of the caged roller type. If wear is evident in the form of play or if roughness is indicated when they are rotated, replacement is necessary. Always check for wear when the old oil has been washed out of the assembly. Whilst it is possible to remove the outer bearings at each end of the crankshaft, it is probable that the centre bearings are also worn. Here again, it will be necessary to obtain a replacement crankshaft assembly.

18.1 The main bearings are of the caged roller type

6 Failure of both the big end bearings and the main bearings may not necessarily occur as the result of high mileage covered. If the machine is used infrequently or for a succession of short journeys, it is possible that condensation within the engine may cause premature bearing failure. The condition of the flywheels is usually the best guide. When condensation troubles occur, the flywheels will tend to rust and become discoloured.

19 Oil seals: examination and replacement

1 Oil seal failure is difficult to define precisely. Usually it takes the form of oil showing on the outside of the machine, and there is nothing worse than those unsightly patches of oil on the ground where the machine has been standing. One of the most crucial places to look for an oil leak is behind the gearbox final drive sprocket. The seal and 'O' ring that fits on the shaft should be renewed if there is any sign of a leak.
2 Another seal to watch is the clutch pushrod seal that fits into the crankcase in front of the gearbox sprocket. This seal can be replaced from the outside.

20 Cylinder block: examination and renovation

1 The usual indication of badly worn cylinder bores and pistons is excessive smoking from the exhausts. This usually takes the form of a blue haze tending to develop into a white haze as the wear becomes more pronounced.
2 The other indication is piston slap, a form of a metallic rattle which occurs when there is little load on the engine. If the top of the bore is examined carefully, it will be found that there is a ridge on the thrust side, the depth of which will vary according to the rate of wear which has taken place. This marks the limit of travel of the top piston ring.
3 Measure the bore diameter just below the ridge using an internal micrometer, or a dial gauge. Compare the reading you obtain with the reading at the bottom of the cylinder bore, which has not been subjected to any piston wear. If the difference in readings exceeds 0.005 mm (0.002 in.) the cylinder block will have to be bored and honed, and fitted with the required oversize pistons.
4 If a measuring instrument is not available, the amount of cylinder bore wear can be measured by inserting the piston (without rings) so that it is approximately ¾ inch from the top of the bore. If it is possible to insert a 0.005 inch feeler gauge between the piston and cylinder wall on the thrust side of the piston, remedial action must be taken.
5 Kawasaki supply pistons in two oversizes: 0.020 inch (0.5

mm) and 0.040 inch (1.0 mm). If boring in excess of 1.0 mm becomes necessary, the cylinder block must be renewed since new liners are not available from Kawasaki.
6 Make sure the external cooling fins of the cylinder block are free from oil and road dirt, as this can prevent the free flow of air over the engine and cause overheating problems.

21 Pistons and piston rings: examination and renovation

1 If a rebore becomes necessary, the existing pistons and piston rings can be disregarded because they will have to be replaced by their new oversizes.
2 Remove all traces of carbon from the piston crowns, using a blunt ended scraper to avoid scratching the surface. Finish off by polishing the crowns of each piston with metal polish, so that carbon will not adhere so rapidly in the future. Never use emery cloth on the soft aluminium.
3 Piston wear usually occurs at the skirt or lower end of the piston and takes the form of vertical streaks or score marks on the thrust side of the piston. Damage of this nature will necessitate renewal.
4 The piston ring grooves may become enlarged in use, allowing the rings to have a greater side float. If the clearance exceeds 0.005 inch the pistons are due for replacement.
5 To measure the end gap, insert each piston ring into its cylinder bore, using the crown of the bare piston to locate it about 1 inch from the top of the bore. Make sure it is square in the bore and insert a feeler gauge in the end gap. If the end gap exceeds 0.028 inch (0.7 mm) the ring must be renewed. The standard gap is 0.008 - 0.016 inch (0.2 - 0.4 mm).

When refitting new piston rings, it is also necessary to check the end gap. If there is insufficient clearance, the rings will break up in the bore whilst the engine is running and cause extensive damage. The ring gap may be increased by filing the ends of the rings with a fine file.

The ring should be supported on the end as much as possible to avoid breakage when filing, and should be filed square with the end. Remove only a small amount of metal at a time and keep rechecking the clearance in the bore.

22 Cylinder head: examination and renovation

1 Remove all traces of carbon from the cylinder head using a blunt ended scraper (the round end of an old steel rule will do). Finish by polishing with metal polish to give a smooth shiny surface. This will aid gas flow and will also prevent carbon from adhering so firmly in the future.
2 Check the condition of the spark plug hole threads. If the threads are worn or crossed they can be reclaimed by a Helicoil insert. Most motorcycle dealers operate this service which is very simple, cheap, and effective.
3 Clean the cylinder head fins with a wire brush, to prevent overheating, through dirt blocking the fins.
4 Lay the cylinder head on a sheet of ¼ inch plate glass to check for distortion. Aluminium alloy cylinder heads distort very easily, especially if the cylinder head bolts are tightened down unevenly. If the amount of distortion is only slight, it is permissible to rub the head down until it is flat once again by wrapping a sheet of very fine emery cloth around the plate glass base and rubbing with a rotary motion.
5 If the cylinder head is distorted badly (one way of determining this is if the cylinder head gaskets have a tendency to keep blowing), the head will have to be machined by a competent engineer experienced in this type of work. This will, of course, raise the compression of the engine, and if too much is removed can adversely affect the performance of the engine. If there is risk of this happening, the only remedy is a new replacement cylinder head.

21.2 Use a blunt scraper to remove carbon

23.1 Keep bushes in original locations

23.1A Using the spring compressor, to remove valve

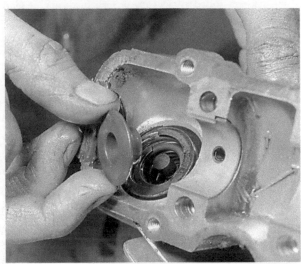
23.1B Take out spring caps

23.1C Note seal round outside of guides

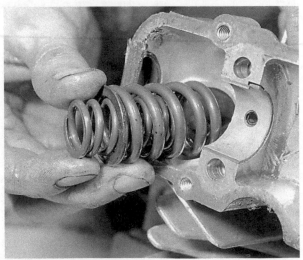
23.2 Remove valve springs

23 Valves, valve seats and valve guides: examination and renovation

1   Remove the valve tappets and shims, keeping them separate for installation in their original locations. Compress the valve springs with a valve spring compressor, and remove the split valve collets, also the oil seals from the valve guides, as it is best to renew these latter components.

2   Remove the valves and springs, making sure to keep to the locations during assembly. Inspect the valves for wear, overheating or burning, and replace them as necessary. Measure the head of the valve using vernier calipers. The standard measurement of the head is: 0.034 - 0.020 inch (0.85 - 1.15 mm). Replace when the measurement is below 0.020 inch (0.5 mm). Carefully inspect the face of the valve for pitting. This can sometimes be taken out by using a coarse grinding paste first and finishing off with a fine grinding paste, using a suction grinder with a rotary action, keep lifting the suction grinder to distribute the paste on the valve face evenly. The valve surface angle is 45°.

3   Carefully inspect the valve seat for pitting, if this is slight it may be possible to grind this out by hand, using the suction grinder and valve technique, as in the preceding paragraph. If the pitting is too bad, the seat will require recutting with a seat cutter.

4   Measure the bore of each valve guide in at least four places using a small bore gauge and micrometer. The standard measurement for the guide (internal diameter) is 0.2756 - 0.2762 inch (7.000 - 7.015 mm). If the measurement exceeds 0.280 inch (7.10 mm) the guide should be replaced with a new one.

5   If a small bore gauge and micrometer are not available, insert a new valve into the guide, and set a dial gauge against the valve stem. Gently move the valve back and forth in the guide and measure the travel of the valve in each direction. The guide will have to be renewed if the clearance between the valve and guide exceeds 0.004 inch (0.10 mm) inlet and exhaust.

6   To install new valve guides, first fit the circlip into the groove round the guide then heat the area around the guide hole to 250° - 300°F (120° - 150°C) and drive the guide in from the top of the cylinder head using a suitable drift, until the circlip reaches its seat. Use a 7 mm reamer to ream the guides. This must be done even if the old guides are used. Note: always rotate the reamer to the right, and keep rotating it until it is withdrawn.

7   Inspect the valve springs for tilt or a collapsed condition. Measure the free length of the springs with a vernier gauge and renew the springs as a complete set if there is any variation in length between the springs. The measurements are as follows:

Inner spring:	*1.42 inch*	*(36.0 mm)*	*Standard*
Inner spring:	*1.38 inch*	*(35.0 mm)*	*Wear limit*
Outer spring:	*1.55 inch*	*(39.3 mm)*	*Standard*
Outer spring:	*1.50 inch*	*(38.0 mm)*	*Wear limit*

The valve springs should always be replaced as a complete set, to ensure maximum performance.

24 Camshafts, tappets and camshaft sprockets: examination and renovation

1   Inspect the cams for signs of wear such as scored lobes, scuffing, or indentation. The cams should have a smooth surface. The complete camshaft must be replaced if any lobes are worn or indented, through lubrication failure etc.

2   Inspect the valve tappets for damage or fracture, also the shims. Measure the shims with a micrometer gauge. Note: The type of valve tappet used on all machines up to engine number Z1.E.08979 has caused some fracturing of the shims. These tappets were of the concave type, whereas the new type tappets are level. It is advisable to use new shims when replacing these tappets, as the old ones that have been in use may already have started to fracture.

3   Examine the camshaft chain sprockets for worn, broken or chipped teeth. This is important, to keep the cam chain at the

23.2A Take out valves

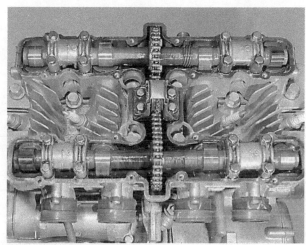
24.1 Inspect cams for wear

24.2 This shim has fractured

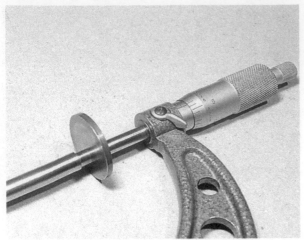

24.2A Measuring shim with a micrometer

24.3 Inspect sprockets for worn teeth

24.3A A broken tensioner, caused through a slack chain

24.5 Remove tensioner body, to examine rubber pad

correct tension at all times. Indeed, the machine featured in this manual had the cam chain tensioner wrongly adjusted, which in turn had broken the cam chain guide through the slackness of the cam chain in operation.

4   A worn camshaft chain will cause a distinctive rattle that will not disappear, even when the tensioner is readjusted. Some indication of the amount of wear that has taken place is given if the chain is held at both ends and twisted sideways. A badly worn chain will bend into a pronounced arc. If there is any doubt about a chain, it should be renewed without question. If a chain breakage occurs, serious engine damage will result.

5   Examine the rubber pad on the end of the chain tensioner push bar. If it is worn or damaged, it should be replaced. The push bar is a push fit into the holder.

25 Clutch and rear chain oil pump: examination and renovation

1   The clutch assembly contains eight friction plates and seven steel plates. The friction plates are steel with bonded cork. The clutch housing incorporates rubber dampers and springs, these are to reduce clutch vibration and snatch. The clutch release outer operating worm gear is made of nylon, and the inner gear is made of steel. There is a clutch adjusting screw located in the

inner release gear which, when turned, adjusts the clutch pushrod clearance. When the clutch needs adjustment, always get this adjustment correct first, before finishing with the operating cable.

2   Check the condition of the clutch drive, to ensure none of the teeth are chipped, broken or badly worn. Give the plain and friction plates a wash with a petrol/paraffin mix and check that they are not buckled or distorted. Remove all traces of friction dust, otherwise a gradual build up of this material will occur and affect clutch action.

3   Visual inspection will show whether the tongues of the clutch plates have become burred and whether corresponding indentations have formed in the slots with which they engage. Burrs in the clutch housing should be removed with a file, also the plates, provided the indentations are not too great.

4   Check the thickness of the friction plates. The standard thickness is: 0.146 - 0.154 inch (3.7 - 3.9 mm). The wear limit is 0.134 inch (3.4 mm).

5   Inspect the clutch plates for warpage by placing them on a sheet of glass or similar flat surface. Place a feeler gauge underneath the plate and the surface. The standard tolerable warpage for the friction plates is under 0.006 inch (0.15 mm) with a service limit of 0.012 inch (0.30 mm). Plain steel plates have a standard tolerable warpage of under 0.008 inch (0.20 mm) with

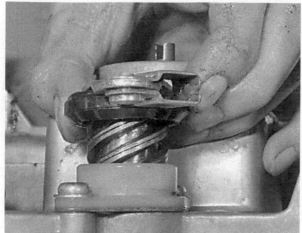

25.1 The clutch operating worm

25.3 Look for burrs in clutch housing

25.9 Clutch adjustment and rear chain oil pump

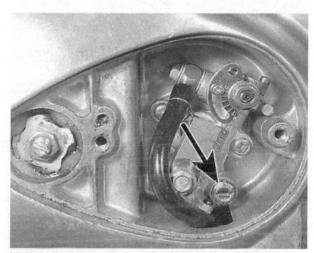

25.10 The bleeder screw in chain oil pump (Z1 model only)

25.11 Groove in shaft aligns in gearbox shaft

a service limit of 0.016 inch (0.40 mm).

6 Inspect the clutch springs by standing them together on a flat surface. If they vary in height, the complete set should be renewed.

7 Roll the clutch pushrod on a sheet of glass to check for trueness. Also make sure the clutch operating worm gear is greased for easy action.

8 Make sure that the needle roller bearing is free and not damaged. The whole clutch assembly rotates on this bearing and there should be no undue wobble.

9 The early Z.1. models are fitted with a rear chain oil pump, driven by the layshaft and fed from a separate oil tank located behind the left hand cover under the dualseat. The oil pump is located in the outer chain cover fitted over the rear chain final drive sprocket. To gain access, unscrew the two crosshead screws that hold the pump to the cover, and remove the pump. The 'O' rings and the check valve are the only parts that are likely to wear, and consequently are the only available replacement parts. If any other parts of the pump are worn or damaged, the entire pump assembly will have to be replaced.

10 Check that the check valve will pass oil only in one direction and that it is not clogged. The valve can be cleaned with petrol or solvent in a syringe. Whenever the oil pump is dismantled, or if the pump runs dry while the pump is in operation, air will

enter the pump. The pump must then be bled of air by turning the bleeder bolt until oil flows from the hole and then re-tightening it.

11 Lubricate all the parts of the pump when reassembling, then rotate the rear wheel until the pin in the layshaft (located through the gearbox sprocket) is brought into alignment with the groove in the oil pump shaft. This is essential for correct installation, otherwise the pump will not seat correctly. Bleed the pump and make sure the level·in the oil tank is correct.

26 Gearbox components: examination

1 Give all the gearbox components a close visual examination for signs of wear or damage, such as chipped or broken teeth, worn dogs or worn splines and bent selector arms. If the machine has shown a tendency to jump out of gear, look especially for worn dogs on the back of the gear pinions or wear in the selector tracks of the gear selector drum. In the former case, wear will be evident in the form of rounded corners or even a wedge - shaped profile in an extreme case. The corners of the selector drum tracks will wear first, all such wear is evident by the appearance of a brightly polished surface.

2 In the selector arms wear usually occurs across the fork that engages with the sliding gear pinions, causing a certain amount of sloppiness in the gear change movement. A bent selector will immediately be obvious, especially if overheating has blued the surface.

3 All gearbox components that prove faulty will have to be replaced with new ones, as there is no satisfactory method of reclaiming them.

27 Reassembling the cylinder head

1 Commence reassembly by replacing the eight valves (four inlet and four exhaust). Do not omit to lubricate the valve stems with engine oil before inserting them into the guides, and make sure that the valve collets are seated correctly. A light tap on the end of the valve stem, after assembly, is a good check. Make sure new oil seals are fitted to the valve guides.

2 Replace the bearing halves for the camshafts, making sure that their notches locate into the cylinder head casting, and cylinder head cover, then install the tappets and shims to their correct locations. Any shims or tappets that show signs of fracture (inspect under a magnifying glass) must be replaced with new ones. Shims are supplied in different thicknesses to facilitate valve clearance adjustment on an individual basis.

28 Engine and gearbox reassembly: general

1 All the engine and gearbox components should be cleaned thoroughly before they are replaced, and be placed close to the working area.

2 The castings and covers that have all had gaskets attached should be cleaned, preferably with a clean rag soaked with methylated spirits. This will act as a solvent for the gasket cement. This is a far better method than reverting to scraping, with the risk of damage to the mating surfaces.

3 Have available all the necessary tools and a can filled with clean engine oil. Make sure all the new gaskets and any replace-ment parts required are available and to hand. There is nothing more frustrating than having to stop in the middle of a reassembly sequence because a vital gasket or part has been overlooked. Always use genuine 'Kawasaki' parts.

4 Make sure the working area is well lit and clean, with plenty of space, as the Kawasaki engine is very big and fairly heavy, although easy to work on. Refer to the torque and clearance settings wherever they are given. Many of the smaller bolts are easily sheared if overtightened. Always use the correct size screwdriver bit for the crosshead screws and NEVER an ordinary flat blade screwdriver or worse still a punch. Any screws having heads that are badly damaged will prove almost impossible to

remove on the next occasion.

29 Engine and gearbox reassembly: replacing the crankshaft

1 Invert the upper crankcase and place it on the workbench. Commence reassembly by lowering the crankshaft assembly into position, taking care to locate all the holes in the outer tracks of the main bearings with the corresponding dowels located in the crankcase. Make sure the cam drive chain is looped over the crankshaft at this stage. Feed the connecting rods through the apertures in the crankcase and once the crankshaft is in position, rotate it to make sure all the main bearings revolve freely, and that the six main bearings are located correctly with the six dowels.

2 Refit the crankshaft centre bearing cap into position. The bearing cap is bored in line with the crankcase, and must be installed with the arrow on the cap pointing towards the front of the engine. Secure the four bolts with 18 ft lbs (2.5 kg m) torque in the numbered sequence shown.

27.2 Fit bushes into correct location

29.1 The crankshaft and camchain. Note white marks for timing purposes

29.1A Lower the crankshaft into the crankcase

29.1B Location holes in main bearings mate up with dowels in crankcase

29.1C Oil holes must also align with holes in case

29.2 Arrow on cap must point to front of engine

30.2 Fit seegar circlip with circlip pliers

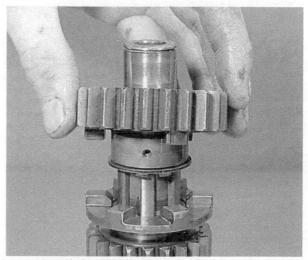

30.2A Note bush inside fifth gear 23T

30 Engine and gearbox reassembly: rebuilding the gearbox

1 With the upper crankcase still inverted on the bench and with the crankshaft assembly installed, start to rebuild the gearbox by assembling the mainshaft gear cluster.

2 Replace the 4th gear with bush 21T. The splined washer, with the external seegar circlip follows and then the 3rd gear (21T) seegar circlip, splined washer, the bush that fits inside the 5th gear and the 5th gear (23T). Next replace the small 2nd gear (16T), the two spacers, the needle roller bearing together with the shim and seegar circlip, and the end bush. Place the bearing on the opposite end of the shaft, and make sure the set rings that locate the bearings are in position.

3 To rebuilt the layshaft cluster replace the shim behind the large 1st gear (38T). This is the only gear with the six large holes bored in it. Follow up with the two spacers, the needle roller bearing, the shim, seegar circlip and the outer bush. Next replace the three steel balls inside the 4th gear (29T) and refit the gear to the shaft. The reason for the steel balls inside the gear pinion is to facilitate the selection of neutral when changing from 1st gear.

4 Replace the distance washer, the 3rd gear (35T) the splined washer, seegar circlip and the 5th gear (28T). Then replace the seegar circlip, the distance splined washer and the 2nd gear bush together with the 2nd gear (35T). This is the large gear with the six large elongated holes. Next replace the distance washer and the gearbox main bearing. Place the set ring in position.

5 The layshaft cluster can now be assembled into the inverted upper crankcase making sure to locate the main bearing with set ring into the groove and the dowel in the crankcase with the hole in the bush at the opposite end of the shaft.

6 Replace the mainshaft cluster, again making sure that the main bearing engages with the set ring in the crankcase, also that the crankcase dowel engages with the bush at the opposite end of the shaft.

7 The kickstart shaft assembly can now be replaced, making sure to position the ratchet onto the stop, which is located on the crankcase.

8 The clutch housing with needle roller bearing, and thrust washer, positioned on the shaft behind the housing, can now be replaced. This **must** be carried out before the crankcase halves are joined.

9 Replace the gear selector drum with the selector fork into the lower crankcase half, and locate the drum with the indent pin. Bend over the tab washer on the pin. The indent pin locates in the end track of the drum and is spring loaded.

10 Locate the selector forks with the sliding 4th and 5th gear pinions, then enter the selector fork shaft through the side of the crankcase. When assembled with the forks in the correct position, fit the external circlip over the end of the rod to locate the rod.

11 Before joining the crankcase halves make sure the 'O' ring is located in the centre of the crankcase. This seals the main oil passage for the whole of the engine. Thoroughly clean the crankcase mating surfaces, and apply a liquid jointing cement to the surface of the lower half. Secure the eight 8 mm bolts in the lower crankcase half to 18 ft lbs (2.5 kg m). Tighten evenly following the order of the numbers stamped on the lower crankcase half. The threads of bolts number 6 and 8 should be coated with sealer, to prevent oil seepage.

12 Now secure the twenty-two 6 mm bolts to a torque of 70 inch lbs or (0.8 kg m) and coat the thread of the remaining bolt with a liquid sealant.

13 Install the oil pump with the 'O' ring inserted in the crankcase, the pump is located by two dowels. Check that the oil pump drive gear meshes correctly with the crankshaft gear when installing the oil pump. Tighten the two mounting bolts to a torque setting of 70 inch lbs (0.8 kg m).

14 Replace the 'O' ring on the layshaft before installing the transmission cover, to prevent oil leaks. Use a guide to protect the oil seal when replacing the transmission inner cover, and

30.2B Replace second gear 16T

30.2C Fit spacers last

30.2D Set rings engage in grooves

30.3 Replacing layshaft gears

30.4 Fit splined washer before third gear

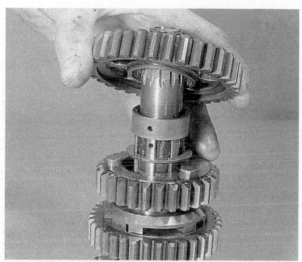
30.4A Bush inside second gear

30.5 Layshaft cluster assembled

30.7 Replace assembled kickstart shaft

30.7A Make sure ratchet engages with stop

30.8 Gear clusters and clutch in position

30.9 Make sure groove in drum is in line with bolt hole

30.9A Fit locktab on bolt

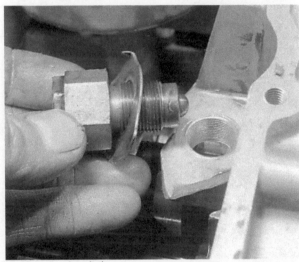

30.9B and replace bolt

30.9C Bend over lock tab

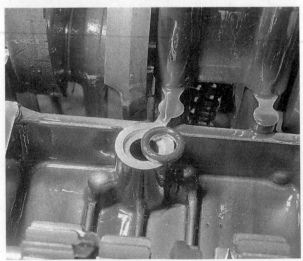

30.11 'O' ring fits in crankcase

30.11A Clutch must be replaced before crankcases are joined

30.11B Make sure surfaces are clean

30.11C Note bolts in crankcase well

30.13 Install 'O' ring in oilway before fitting oil pump

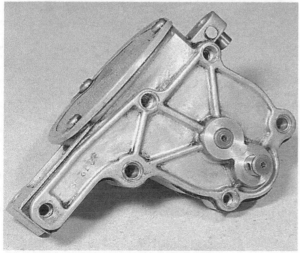
30.13A The pump is located with two dowels

30.14 Replacing inner transmission cover

30.14A Don't omit distance collar

30.15 Refit 91 mm 'O' ring with a new gasket to sump cover

30.15A Make sure oil filter is clean

31.1 Two screws secure breather base

install the distance collar over the layshaft when the cover is in place, otherwise the seal will be damaged.

15 Replace the oil sump cover when the 91 mm 'O' ring is in place and use a new gasket. When the sump is bolted up, fit a new oil filter element to the filter unit, not omitting the spring and 'O' ring. When the sump bolts have been tightened up and the drain plug secured, the engine unit can be turned over to its correct position, with the crankcase apertures facing upwards.

31 Engine and gearbox: rebuilding the clutch, alternator and contact breaker

1 Replace the breather assembly on top of the upper crankcase with the two crosshead screws, and the polished alloy cover with the entire fixing and pipe for the breather tube. The oil pressure switch can now be replaced. Make sure the two 'O' rings are in position first. Replace the rubber cap.

2 Rotate the kickstart shaft all the way round to the right before hooking the return spring into the hole in the shaft, this gives the spring the correct tension. The kickstart cover can now be replaced with the four screws.

3 Replace the detent lever with the bolt and locate the return spring with the pin. The detent lever engages behind the shroud on the selector drum.

4 The footchange lever shaft complete with the selector pawls and spring can now be inserted into the crankcase, the sprocket distance collar and transmission cover can also be fitted.

5 The rotor and starter clutch gear can now be fitted. Make sure the rollers are free to move in the starter clutch. The idler gear can be replaced, after checking to ensure the starter clutch gear meshes correctly with the idler gear, and that the thrust washer is fitted behind the starter gear with its chamfered inner edge facing inwards. Inspect the rotor needle roller bearing to ensure free running. The Woodruff key should be a tight fit in the taper shaft. To tighten the rotor securing bolt, place a round bar through a connecting rod small end and hold tight on the downward stroke of the connecting rod. The bar will lock the crankshaft assembly, enabling the rotor bolt to be tightened.

6 The rotor cover with coils attached can now be fitted, making sure that the wiring is located through the grommet in the side of the cover.

7 Fit the starter motor with a little oil on the 'O' ring to facilitate its assembly into the aperture.

8 Replace the clutch hub centre. Install the distance washer first, then the hub, reset the outside washer (so named) with

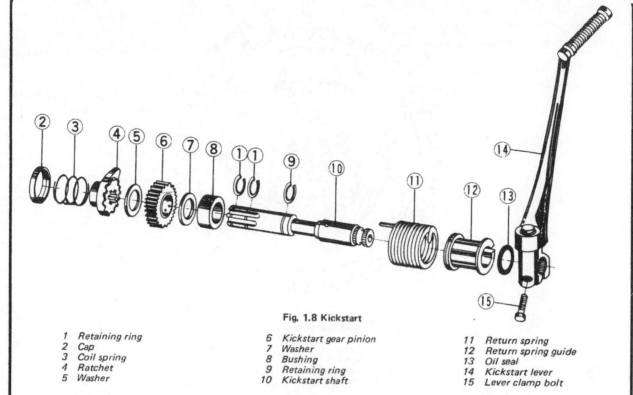

Fig. 1.8 Kickstart

1	Retaining ring	6	Kickstart gear pinion	11	Return spring
2	Cap	7	Washer	12	Return spring guide
3	Coil spring	8	Bushing	13	Oil seal
4	Ratchet	9	Retaining ring	14	Kickstart lever
5	Washer	10	Kickstart shaft	15	Lever clamp bolt

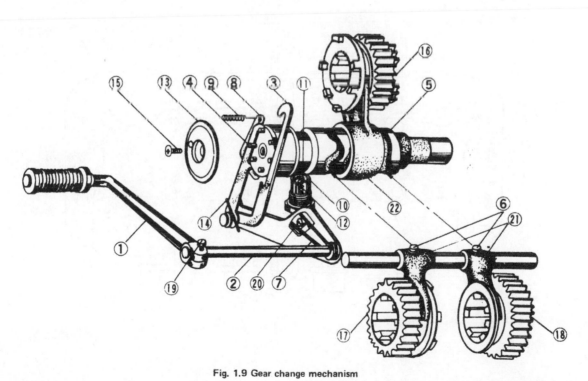

Fig. 1.9 Gear change mechanism

1	Gear change deal	7	Return spring	12	Spring	17	Layshaft 5th gear
2	Shaft	8	Detent arm	13	Selector drum	18	Layshaft 4th gear
3	Selector pawl	9	Detent arm spring		pin holder	19	Bolt
4	Selector drum pin - 6 off	10	Selector drum	14	Pawl spring	20	Return spring pin
5	Selector drum		positioning pin	15	Screw	21	Selector fork
6	Selector fork pin	11	Neutral detent pin	16	Mainshaft 3rd gear	22	Selector fork

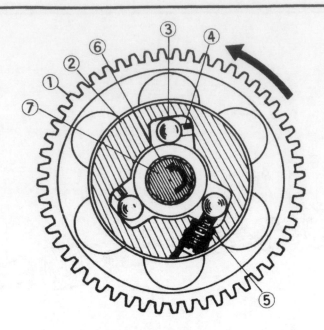

Fig. 1.10 Starter clutch

1	Gear	5	Spring - 3 off
2	Clutch body	6	Needle roller
3	Roller - 3 off		bearing
4	Spring cap	7	Crankshaft

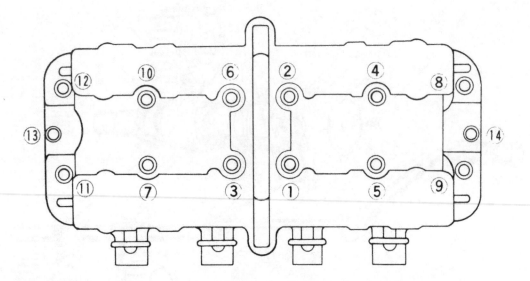

Fig. 1.11 Cylinder head nut tightening sequence

31.1A Replacing the crankcase breather

31.1B Make sure 'O' rings are in position

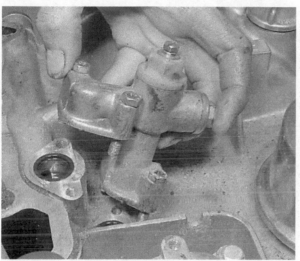
31.1C When replacing oil switch body

31.1D Replace protective rubber cap on switch

31.2 Preload kickstart shaft. when replacing spring

31.2A Use new gasket when replacing cover

'outside' facing outwards, and then the Simmonds type clutch centre locknut. Tighten the nut with a socket spanner and bar. If no clutch tool is available, it is possible to jam the hub with a large tyre lever and box spanner to give just enough pressure to tighten the nut. Great care is necessary, however, otherwise damaged hub splines will result.

9 Grease and replace the clutch pushrod, the steel ball, and the mushroom pressure piece. Replace the clutch plates starting with a friction plate first and alternating with a plain plate and then a friction plate, to finish with a friction plate and lastly the pressure end plate. Replace the clutch pressure springs, the spring washers and the five bolts. The bolts can then be fully tightened.

10 Fit a new gasket to the clutch cover base, and fit the clutch cover, making sure that the dowels in the crankcase register with the holes in the cover. Secure the cover with the nine crosshead screws.

11 Refit the contact breaker case cover with a new gasket, first making sure the oil seal is in good condition, otherwise oil will get through to the contact breaker assembly.

12 Assemble the advance and retard mechanism making sure the pin on the end of the crankshaft engages with the slot in the advance mechanism. Refit the cam and align the mark on the cam with the mark on the advance mechanism body.

13 Mount the contact breaker assembly base plate with the points and condensers in situ, and secure with the three crosshead screws. Do not put the contact breaker cover on yet, since it will be necessary to adjust the points gap and time the ignition at a later stage.

31.3 Fitting detent lever in selector drum

---

32 Engine and gearbox reassembly: replacing the pistons and cylinder block

---

1 Before replacing the pistons, pad the mouths of the crankcase with rag in order to prevent any displaced component from accidentally dropping into the crankcase.

2 Fit the pistons in their original order with the arrow on the piston crown pointing toward the front of the engine.

3 If the gudgeon pins are a tight fit, first warm the pistons to expand the metal. Oil the gudgeon pins and small end bearing surfaces, also the piston bosses, before fitting the pistons.

4 Always use new circlips never the originals. Always check that the circlips are located properly in their grooves in the piston boss. A displaced circlip will cause severe damage to the cylinder bore, and possibly an engine seizure.

5 Place a new cylinder base gasket (dry) over the crankcase mouth. Note the extra hole in the gasket must be on the right hand side of the engine. Refit the bottom guide roller. Now place the cylinder block over the cylinder studs (make sure the four 'O' rings are fitted to the base of the cylinders), support the cylinder block whilst the camshaft chain is threaded through the tunnel between the bores. This task is best achieved by using a piece of stiff wire to hook the chain through, and pull up through the tunnel. The chain must engage with the crankshaft drive sprocket.

6 The cylinder bores have a generous lead in for the pistons at the bottom, and although it is an advantage on a large engine such as this to use the special Kawasaki ring compressor, in the absence of this, it is possible to gently lead the pistons into the bores, working across from one side. Great care has to be taken NOT to put too much pressure on the fitted piston rings. Do not omit to fit the two dowels to the front outside studs. When the pistons have finally engaged, remove the rag padding from the crankcase mouths and lower the cylinder block still further until it seats firmly on the base gasket.

7 Take care to anchor the camshaft chain throughout this operation to save the chain dropping down into the crankcase. The two idler assembly sprockets that guide the cam chain, can now be replaced, with their shafts and rubbers, into the top of the cylinder block, and the chain guide screwed into position.

31.3A Footchange shaft with circlip being inserted

31.4 Gear selector pawls in position

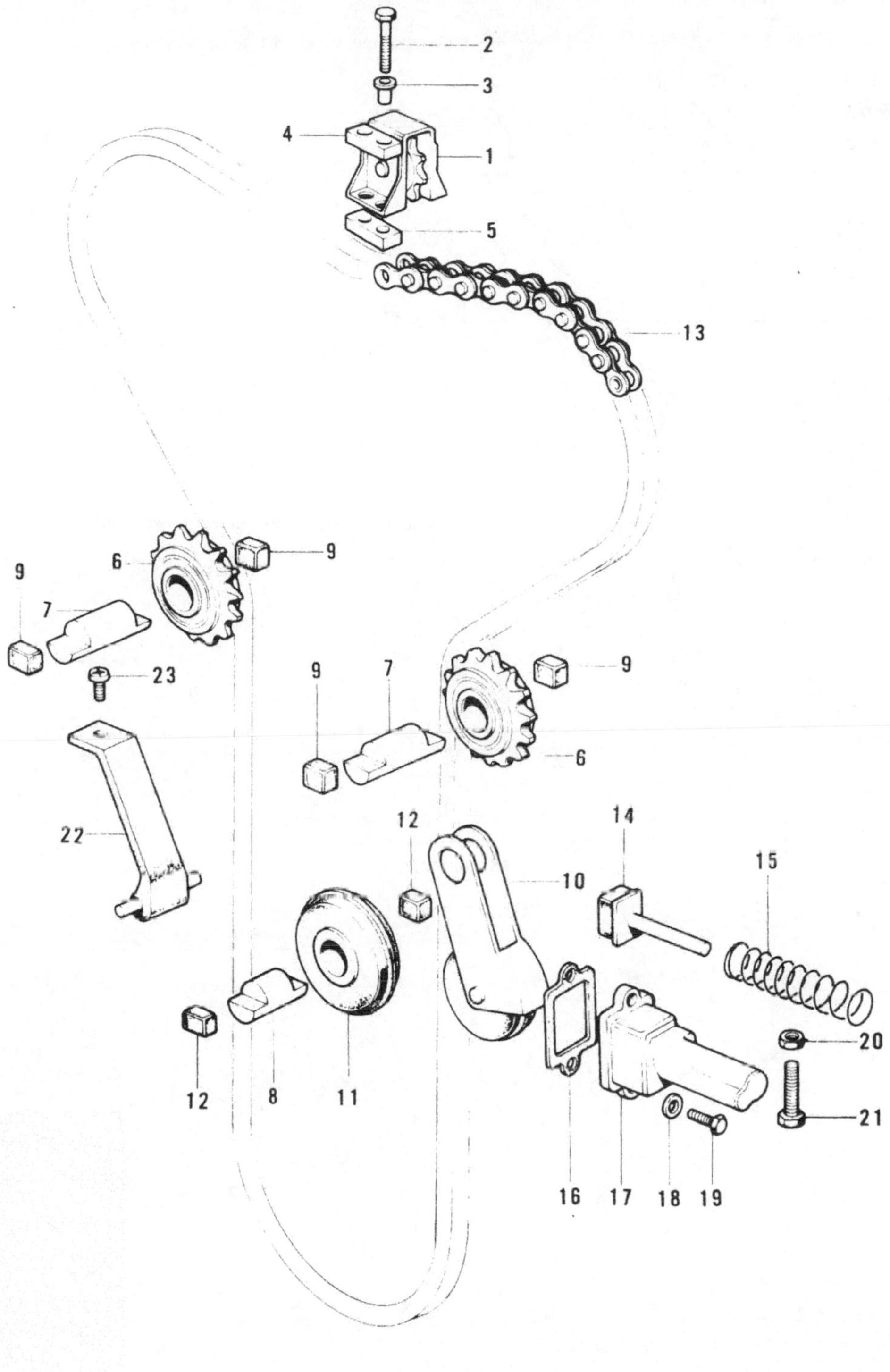

Fig. 1.12 Cam chain and tensioner

1 Cam chain idler assembly	sprocket - 2 off	11 Roller	17 Tensioner holder
2 Bolt - 4 off	7 Idler sprocket shaft - 2 off	12 Roller rubber - 2 off	18 Washer - 2 off
3 Idler collar - 4 off	8 Guide roller shaft	13 Cam chain	19 Bolt - 2 off
4 Idler rubber - 2 off	9 Rubber guide roller - 4 off	14 Pushrod assembly	20 Nut
5 Idler rubber - 2 off	10 Cam chain tensioner	15 Tensioner spring	21 Bolt
6 Cam chain idler	assembly	16 Gasket	22 Cam chain guide
			23 Pan head screw

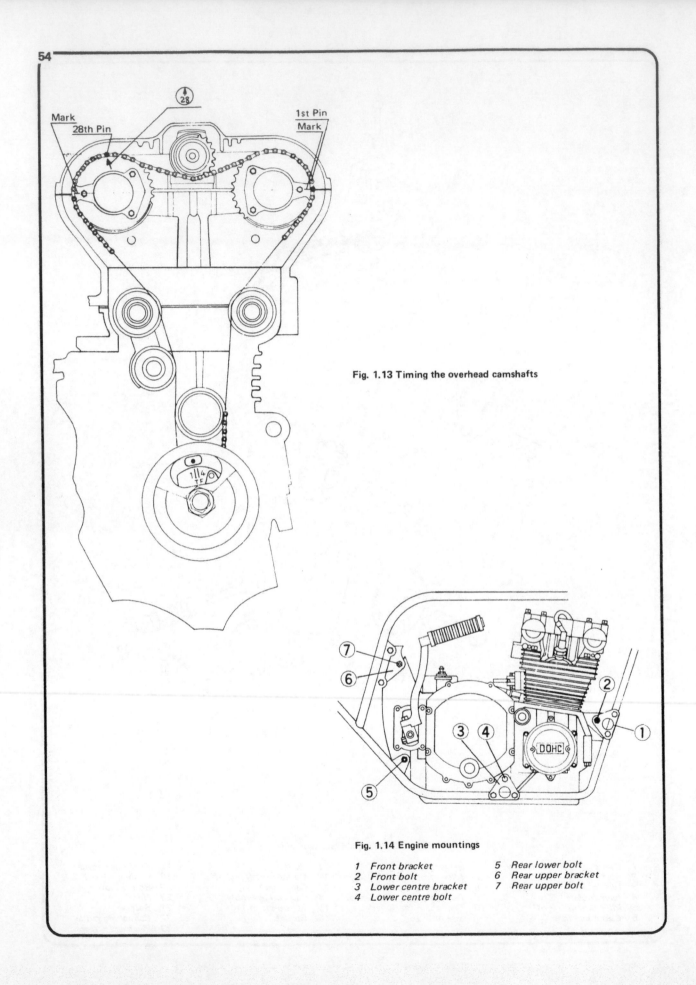

Fig. 1.13 Timing the overhead camshafts

Fig. 1.14 Engine mountings

1 Front bracket	5 Rear lower bolt
2 Front bolt	6 Rear upper bracket
3 Lower centre bracket	7 Rear upper bolt
4 Lower centre bolt	

33 Engine and gearbox reassembly: replacing the cylinder head and camshafts

1 Rotate the crankshaft until the 'T' mark on the advance and retard mechanism is aligned with the timing mark as shown in the accompanying photograph. At this position, number one and four pistons are at top dead centre.

2 Replace the valve tappets and the shims in their original locations and use a new cylinder head gasket to prevent any compression leakage. The side of the gasket with the wider turned over metal edge faces upwards. The cylinder head can now be bolted down, tightening the nuts diagonally to 25 ft lbs. (3.5 kg m)

3 After the cylinder head has been secured, the next operation is to fit the camshafts. Start by fitting the exhaust camshaft first. To fit the camshaft, feed the camshaft through the cam chain, and turn the camshaft so that mark on the sprocket is aligned with the cylinder head surface, as shown in the accompanying photograph.

4 Now pull the cam chain taut and fit the chain on to the camshaft sprocket. Starting with the next chain link pin above the one that coincides with the sprocket mark, count the pins, until you reach the 28th pin and slide the inlet camshaft into position so that the 28th pin coincides with the number 28 on the rubber portion of the inlet camshaft sprocket.

5 Having assembled the camshafts and replaced the cam chain, the next task is to bolt the camshafts down.

6 Assemble all the split bush halves back into their original positions, and replace the camshaft caps. The caps are machined 'in line' with the cylinder head, and so it is very important that they are replaced with the number on the cap corresponding to the number on the cylinder head. Also the arrows marked on the caps must point to the front.

7 Partially tighten the left hand caps first, to seat the camshafts in place. All the bolts can now be fully tightened down to 104 in lbs torque or (1.2 kg m). They should be tightened down in sequence.

8 Make sure all the camshaft bearings and valve tappets are lubricated with clean engine oil. The top chain guide sprocket can now be installed. Adjust the cam chain tension by refitting the tensioner. When this is installed loosen the locknut on the adjuster so that the plunger rod is free to move, rotate the engine slowly a couple of times to make sure the spring loaded tensioner takes up the slack in the chain evenly, and then tighten the bolt first and reset the locknut. It will adjust to the correct tension automatically.

9 To make double sure the timing is right rotate the engine until pistons numbers one and number four are at T.D.C. and check that both the mark on the exhaust camshaft sprocket and the mark on the metal part of the inlet camshaft sprocket are aligned level with the cylinder head surface. This will indicate that the cam timing is correct. CAUTION. Always use a spanner on the large nut on the crankshaft when turning the engine over for timing purposes. DO NOT turn the engine by turning the camshaft sprockets.

34 Engine and gearbox reassembly: adjusting the valve clearances

1 To adjust the valve clearances it is advisable to use the Kawasaki special lifter tool. This consists of a small piece of metal with a turned up edge. When clamped to the side of the cylinder head camshaft cover, it holds down the valve in question and permits the removal and replacement of the shims as the correct adjustment is achieved. The special tool part number is 57001 - 109.

2 Commence adjusting the valves by checking one valve at a time. Turn the crankshaft so that the cam lobe (this is the highest part of the cam) is pointing directly upwards, away from the tappet and insert a .004 inch (0.10 mm) feeler gauge. If the clearance is too small or there is no clearance at all, use the special tool to push down the valve tappet and remove the shim. To use the tool correctly rotate the engine so that the cam lobe

is pushing the tappet down and opening the valve. When the valve is fully open, clamp the tool to the side of the lower cover with the screw provided, and turn the engine so that the cam lobe points upward away from the tappet. This will give access to the tappet shim which can be removed and replaced as required, by using a pair of ordinary eyebrow tweezers, as these are a very thin gauge. There is a notch in the tappet so that the shim can be grasped.

3 Insert a new shim to bring the tappet clearance within the specified limits. Shims are available in sizes from 2.00 to 3.00 mm in increments of .05 mm.

4 If the valve clearance is greater than 0.10 mm (0.004 in.) use a thicker shim to correct to the specified clearance. If the valve clearance is less than 0.05 mm (.002 in.) select a thinner shim. Note: If there is no clearance between the shim and the cam, select a shim which is several sizes smaller and then remeasure the gap.

5 When checking valve clearances always check with the cam lobe pointing upwards directly away from the valve. Checking valve clearances in any other position may result in a false reading.

6 Install the tachometer drive pinion in the front of the cylinder head and screw in the tachometer cable.

7 The camshaft cover can now be replaced and tightened down to a setting of 104 in lbs (1.2 kg m).

8 The spark plugs can now be replaced and the plug leads refitted. The leads are numbered from one to four. The firing order is 1, 2, 4, 3. Number one cylinder for timing purposes is the extreme left hand cylinder.

35 Engine and gearbox reassembly: replacing the carburettors

1 Make sure the holding plate is secured tightly to the four carburettors with the eight countersunk screws before replacing the whole carburettor bank. Also check that the throttle control cable wheel operates and returns freely on the return spring, and that the choke lever operates the chokes of all four carburettors.

2 Secure the carburettors to the intake hoses on the cylinder head by the securing clips fitted round the intake hoses. Make sure these clips are tight, otherwise leakage will occur on the intake side of the carburation and cause irregular running.

3 Channel the four rubber overflow pipes through the retaining band at the rear of the engine, adjacent to the oil filler cap. The engine is now sufficiently complete for installation into the frame.

36 Replacing the engine and gearbox into the frame

1 The task of replacing the engine requires three people, two to lift the engine and one to hold the frame steady while the engine is lowered into position.

2 Lift the complete engine unit into the frame from the right hand side and mount but do not secure the three engine mounting brackets, before inserting the engine bolts. Insert the three long bolts from the left hand side of the machine and fit the two spacers on the rear upper bolt. The long spacer is on the left hand side of the engine and the short spacer on the right hand side. Secure the bracket bolts and brackets to the frame. Refit the air cleaner base stay.

3 Install the final drive sprocket to the engine with the chain already fitted to the sprocket, fit the lockwasher and tighten the locknut to 108 ft lbs (15 kg m) torque.

4 An oil seal guide or a suitable substitute should be used when installing the chain cover and chain oil pump. Make sure that the groove in the oil pump lines up with the slot in the driveshaft. When replacing the cover, great care must be taken to avoid damaging the oil seal when replacing the cover (Z1 models only).

5 Remake the electrical connections from the generator wiring harness by means of the various connectors provided. Thread the starter motor cable through to the starter motor and reconnect the cable with the solenoid at the side of the frame. Attach the

positive lead to the battery terminal, and connect the spark plug leads with the spark plugs by means of the push on covers.

6 Refit the two throttle cables to the wheel type control lever (one cable opens the throttles, and one cable closes them), making sure that the opening cable is fitted to the rear, and the closing cable is fitted to the front of the operating wheel.

7 Replace the air cleaner, which is a push on fit to the carburettor air intakes. Tighten the clips that hold the air cleaner to the intakes, by means of the crosshead screws. Connect the crankcase breather tube, also the clutch cable.

8 Lower the petrol tank into position and reconnect the fuel pipes with the petrol tap.

9 Replace the exhaust pipes and silencers by installing the inside silencers first, and then the outside silencers. Fit new manifold gaskets into the cylinder head. The finned manifold clamp can be used to hold the split collars while installing the exhaust pipes. Secure the silencer connecting hose clamp to prevent exhaust leaks, then tighten the rear silencer mounting bolts first, followed by the bolts at the cylinder head, in that order.

10 Refit the footchange lever and kickstart levers in their original positions as denoted by the punch marks and tighten them on their splines by means of the clamp bolts.

11 Refit the riders footrests and tighten them securely, also the pillion footrests.

12 Check that the crankcase drain plug has been secured, and then refill the engine with the correct amount of engine oil. The level can be checked through the sight window in the clutch cover which should be between the two marks.

13 Refill with SAE 10W/40 engine oil, 1 US gallon including the filter or (4.0 litres). The oil tank fitted beneath the left hand side of the dualseat for the rear chain oil pump should be refilled with SAE 90 gear oil, and the capacity is 0.95 U.S. quart or (0.9 litres). (Z1 model only).

14 Check the ignition timing as described in Chapter 3 and when the timing is correct, replace the contact breaker end cover, also the rotor cover, with their respective crosshead screws.

37 Starting and running the rebuilt engine unit

1 Make sure that all the components are connected correctly. The electrical connectors can only be fitted one way, as the wires are coloured individually. Make sure all the control cables are adjusted correctly. Check that the fuse is in the fuse holder, try all the light switches and turn on the ignition switch. Close the choke lever to start.

2 Switch on the ignition and start the engine by turning it over a few times with the kickstart or the electric starter, bearing in mind that the fuel has to work through the four carburettors. Once the engine starts, run at a fairly brisk tick-over speed to enable the oil to work up to the camshafts and valves.

3 Before taking the machine on the road, check that the brakes are correctly adjusted, with the required level of hydraulic fluid in the handlebar master cylinder.

4 Make sure the rear chain is correctly tensioned to 3/8 inch up and down play. Also that the front forks are filled with the correct amount of oil.

5 Check the exterior of the engine for signs of oil leaks or blowing gaskets. Before taking the machine on the road for the first time, check that all nuts and bolts are tight and nothing has been omitted during the reassembling sequence.

38 Taking the rebuilt machine on the road

1 Any rebuilt engine will take time to settle down, even if the parts have been replaced in their original order. For this reason it is highly advisable to treat the machine gently for the first few miles, so that the oil circulates properly and any new parts have a reasonable chance to bed down.

2 Even greater care is needed if the engine has been rebored or if a new crankshaft and main bearings have been fitted. In the case of a rebore the engine will have to be run-in again as if the machine were new. This means much more use of the gearbox and a restraining hand on the throttle until at least 500 miles have been covered. There is not much point in keeping to a set speed limit; the main consideration is to keep a light load on the engine and to gradually work up the performance until the 500 mile mark is reached. As a general guide, it is inadvisable to exceed 4,000 rpm during the first 500 miles and 5,000 rpm for the next 500 miles. These periods are the same as for a rebored engine or one fitted with a new crankshaft. Experience is the best guide since it is easy to tell when the engine is running freely.

3 If at any time the oil feed shows sign of failure, stop the engine immediately and investigate the cause. If the engine is run without oil, even for a short period, irreparable engine damage is inevitable.

31.5 Make sure starter clutch rollers are free to move

31.5A Thrust washer fits behind rotor (chamfered inner edge facing inwards)

31.5B Next fit the needle roller

31.5C Replace idler gear

31.5D and rotor

31.6 Note wiring is located in grommet

31.7 Place a little oil on 'O' ring

31.7A Secure two bolts in starter end bracket

31.8 Fit distance washer first

31.8A Note washer has 'outside' mark

31.8B Replace clutch hub, and nut

31.8C Tighten nut on hub

31.9 Replace clutch rod

31.9A Note steel ball behind mushroom pressure piece

31.9B Replacing the pressure plate

31.9C Fitting clutch springs

31.9D Tighten clutch nuts fully

31.10 Fit new gasket

31.11 Replacing contact breaker base cover

31.12 Pin on shaft aligns auto advance mechanism

31.12A Align marks on cam and body

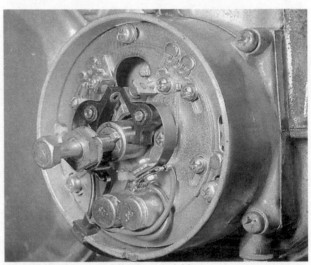
31.13 Replacing contact breaker assembly

32.2 Arrow points towards front of engine

32.5 Fit new base gasket

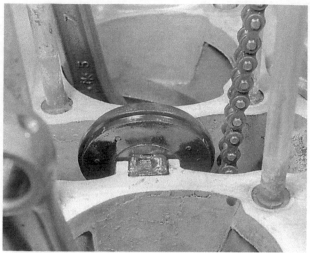
32.5A Bottom guide roller in position

32.6 Rest block over studs when fitting pistons

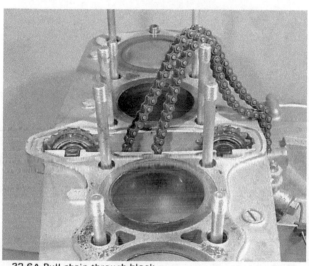
32.6A Pull chain through block

32.6B Settle block over dowels

32.6C Fit gasket metal turned edges up

32.7 Support chain temporarily

33.1 Make sure 'T' mark is aligned with timing mark

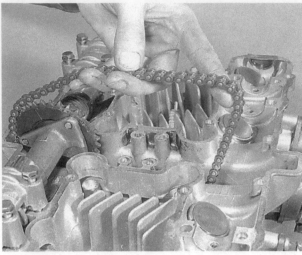

33.3 Fit exhaust camshaft first

33.4 Replace camshaft caps in original positions

33.4A Note No. 28 mark on sprocket

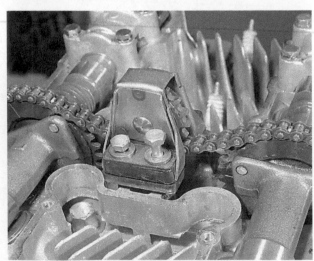

33.8 Installing the top chain guide sprocket

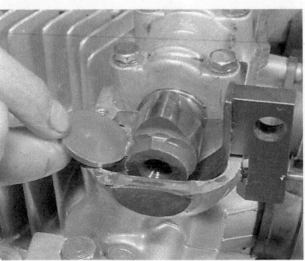

34.1 The special tool clamped into position

34.2 Insert feeler gauge at back of cam

34.5 Make sure cam lobes point up when taking readings

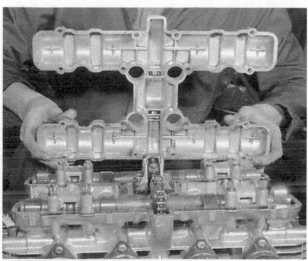

34.7 Replace camshaft cover

35.1 Eight countersunk screws hold plate

35.2 Screws secure carburettors to intake hoses

35.3 Overflow pipes retained by rubber band

36.2 Lift engine in from the right hand side

36.2A Air cleaner base stay

36.2B Stay in position

36.8 Connect up petrol pipes to tap

36.8A Union nut must be tight

36.9 Fit new manifold gaskets

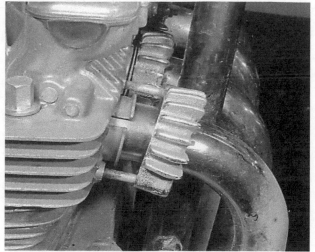
36.9A Split collars fit behind clamp

36.10 Refit levers in original position

36.12 Filling the sump with oil

36.13 Oi level between two marks

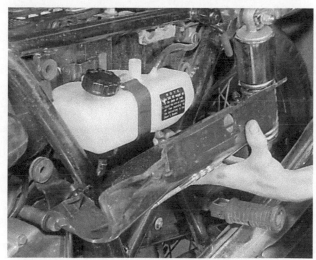
36.13A The rear chain oil tank (Z1 model only)

37.1 Close choke lever for initial start

39 Fault diagnosis : Engine

Symptom	Cause	Remedy
Engine will not start	Defective spark plugs	Remove the plugs and lay on the cylinder head. Check whether spark occurs when engine is on and engine rotated.
	Dirty or closed contact breaker points	Check the condition of the points and whether the points gap is correct.
	Faulty or disconnected condenser	Check whether the points arc when seperated. Replace the condenser if there is evidence of arcing.
Engine runs unevenly	Ignition or fuel system fault	Check each system independently, as though engine will not start.
	Blowing cylinder head gasket	Leak should be evident from oil leakage where gas escapes.
	Incorrect ignition timing	Check accuracy and reset if necessary.
Lack of power	Fault in fuel system or incorrect ignition timing	Check fuel lines or float chambers for sediment. Reset ignition timing.
Heavy oil consumption	cylinder block in need of rebore	Check bore wear, rebore and fit oversize pistons if required.

40 Fault diagnosis: Clutch

Symptom	Cause	Remedy
Engine speed increases as shown by tachmeter but machine does not respond	Clutch slip	Check clutch adjustment for free play, at handlebar lever, check thickness of inserted plates.
Difficulty in engaging gears, gear changes jerky and machine creeps forward when clutch is withdrawn difficulty in selecting neutral	Clutch drag	Check clutch for too much free-play, Check plates for burrs on tongues or drum for indentations. Dress with file if damage not too great.
Clutch operation stiff	Damaged, trapped or frayed control cable	Check cable and replace if necessary. Make sure cable is lubricated and has no sharp bends

41 Fault diagnosis gearbox

Symptom	Cause	Remedy
Difficulty in engaging gears	Selector forks bent Gear clusters not assembled correctly	Replace with new forks. Check gear cluster for arrangement and position of thrust washers.
Machine jumps out of gear.	Worn dogs on the ends of gear pinions	Replace worn pinions.
Gear change lever does not return to original position	Broken return spring	Replace spring
Kickstarter does not return when engine is turned over or started	Broken or wrongly tensioned. Return spring	Replace spring or retension
Kickstarter slips	Ratchet assembly worn	Dismantle engine and replace all worn parts

Chapter 2 Fuel system and lubrication

Contents

General description 1
Petrol tank: removal and replacement 2
Petrol tap and filter: removal, dismantling and replacement 3
Carburettors: description and removal 4
Carburettors: dismantling and reassembly 5
Carburettors: adjustment 6
Carburettors: synchronising 7

Carburettors: settings 8
Air cleaner: dismantling, servicing, reassembly 9
Engine and gearbox lubrication 10
Engine/gearbox lubrication: the oil pump 11
Rear chain lubrication: the chain oil pump 12
Fault diagnosis: fuel system and lubrication 13

Specifications

Fuel tank capacity ... ...   3.5 Imp gallons (16 litres) UK spec.
  4.7 US gallons (18 litres) USA spec.

Engine/gearbox oil capacity ... ...   1.25 Imp gallons/ 4.2 US quart (4 litres)

Rear chain oil tank capacity ... ...   1 Imp quart. (0.95 US quart. (0.9 litre)

Carburettors

		Z1	Z1-B
Make		Mikuni	Mikuni
Type		VM.28.S.C.	—
Main jet		112.5	—
Pilot jet		20	17
Needle jet		P.8	P6
Jet needle		5 J 9-3	—
Needle position		Number 3 Notch	—

1 General description

1 The fuel system comprises a petrol tank, from which petrol is fed by gravity to the carburettors, via a petrol tap that incorporates a filter. There are three positions, ON, RESERVE, and STOP.

2 If the fuel in the tank is too low to be fed to the carburettors in the ON position, turn the lever to the RESERVE position, which provides a further one gallon of reserve supply. The fuel tank cap is the quick action filler type, and incorporates an air vent that must be kept clear for normal fuel flow to occur. If it becomes clogged, it must be blown out with compressed air.

3 A large capacity air cleaner, with a filter element, serves the dual purpose of supplying clean air to the carburettors and effectively silencing the intake roar.

4 The oil content for the engine is contained in a sump at the bottom of the engine. The gearbox is also lubricated from the same source, the whole engine unit being pressure fed by a mechanical oil pump that is driven off the crankshaft gear. The oil pump intake extends into the sump to pump the oil up to the engine. A screen at the pump inlet point prevents foreign matter from entering the pump before it can damage the mechanism. From the pump the oil firstly goes to the oil filter to be cleaned. If the filter becomes clogged, a safety by-pass valve routes the oil around the filter. It next goes through a pipe in which an oil pressure switch is mounted, and through an oil hole in the crankcase, from which point it is sent in three different directions. One direction is to the crankshaft main bearings and crankshaft pins. After lubricating the crankshaft parts, the oil is thrown out by centrifigal force and the spray lands on the cylinder walls and the pistons and piston pins to lubricate those parts. The oil eventually drops down from all these points and accumulates in the bottom of the crankcase sump to be recirculated.

The second passageway for oil from the pump is through the oil passage at each end of the cylinder block and up into the cylinder head. After passing through holes into the camshaft bearings, the oil flows out over the cams and down around the valve tappets to lubricate these areas. The oil returns to the sump via the oil holes at the base of the tappets, and the cam chain tunnel in the centre of the cylinder head and cylinder block.

The third passageway for the oil to flow is to the gearbox bearings where it is pumped to the gearbox main bearing on the mainshaft and also to the bearing on the layshaft. After this the oil drops down back into the oil sump, to be recirculated again through the engine.

2 Petrol tank: removal and replacement

1 The petrol tank fitted to the Z1 models is secured to the frame by means of a short channel that projects from the nose of the tank that engages with a rubber buffer surrounding a pin welded to the frame, immediately behind the steering head. This arrangement is duplicated either side of the nose of the tank and the frame. The rear of the tank is secured by a rubber clip that locates round a lip welded on to the back of the tank. The tank also has two rubber buffers on which it rests at the rear. A petrol tap is fitted with a reserve pipe that is switched over, when the fuel reaches the level of the main petrol pipe. There is no balance pipe.

2 The petrol tank can be removed from the machine without draining the petrol, although the rubber fuel lines to the carburettors will have to be disconnected. The dualseat must be lifted up to release the rubber clip at the rear of the tank then raise the tank at the rear and pull upwards and backwards to pull the tank off of the front rubbers. When replacing the fuel tank, lift at the rear and push down onto the front rubber buffers, then secure the rubber clip at the rear and reconnect the fuel lines.

3 Petrol tap and filter: removal, dismantling and replacement

1 It is not necessary to drain the petrol tank, if the tank is only half or under half full, as the tank can be laid on its side on a clean cloth or soft material (to protect the enamel), so that the petrol tap is uppermost. The petrol tap pipes should be removed, before removing the petrol tap. To remove the tap and filter, first undo the large hexagon union nut next to the tank, the tap body can then be detached complete with the filter. When the filter bowl is removed this will reveal the rubber 'O' ring gasket and the filter gauze. Remove the gauze and clean in petrol. When reassembling the tap, fit a new gasket between the body and the tank and a new rubber 'O' ring to the filter bowl if the old one is noticeably compressed. On no account overtighten the bowl, as it is made of soft metal and the threads will easily strip.

2 There is no necessity to remove either the tap or the petrol tank if only the filter bowl has to be detached for cleaning.

4 Carburettors: description and removal

1 The method of mounting the four carburettors is on a holding plate with eight countersunk screws, the whole bank of carburettors being connected to the intake side of the engine on short induction hoses. The four butterfly type throttle valves are operated by a single shaft, likewise the manually operated choke also has a single shaft operating four levers, one to each carburettor.

2 The throttles are operated by two cables from the handlebar, one to open the throttles and the other to close them. A heavy return spring is incorporated in the throttle return system. When closing the throttles the use of a separate return cable helps to close the throttle more positively. This ensures smooth throttle action.

3 A vacuum gauge fitting is incorporated on each inlet manifold as an aid to balancing manifold pressure, and when used in conjunction with a vacuum gauge array, allows the carburettors to be accurately synchronized.

4 Starting in extreme cold weather is aided by a separate starter system which acts by vacuum pressure and serves in place of a choke. The starter system takes the form of four plunger valves

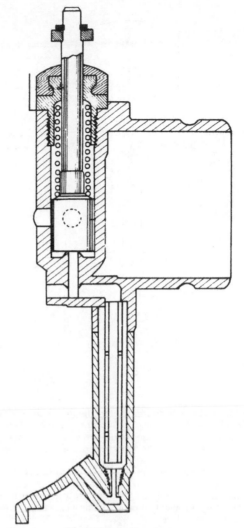

Fig. 2.1 Starter system

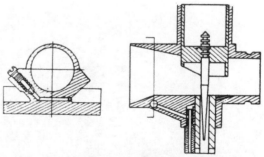

Fig. 2.2 Pilot system

situated at the sides of the carburettors. The four plungers are operated in unison by a lever on the end of the shaft marked 'CHOKE'.

5 With the choke lever raised, and the throttles fully closed, a high intake vacuum is created when the engine is turned over. Fuel flow is metered by the starter jet, and fuel is drawn up

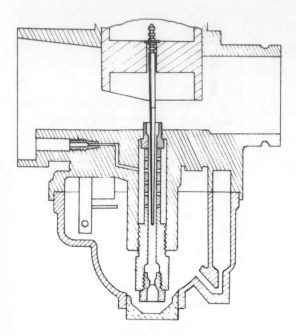

Fig. 2.3 Main system

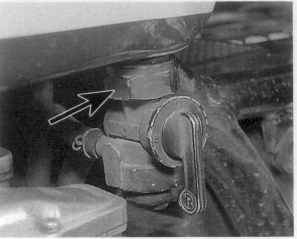

3.1 Undo union nut to remove tap

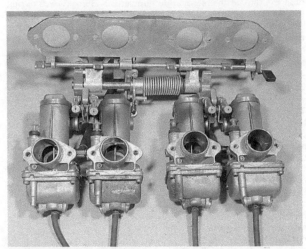

4.1 Mounting plate detached from carburettors

through the starter pipe to the starter plunger chamber where it is atomized. The rich mixture is then sprayed into the carburettor bore where a small fuel/air spray from the pilot system is mixed with it. The final mixture is then delivered to the engine.

It is essential that the starter plungers be fully raised by the choke lever and the starter jet, pipe, and the air bleed hole completely free of any blockage. The throttle must be fully closed so that sufficient vacuum is developed for efficient atomization to take place.

6 The pilot system is made up of the pilot jet, the pilot air screw, and the pilot outlet. It controls carburation from the idle position to approximately one eighth throttle opening. The pilot

4.1A Single shaft connects four throttles

4.2 Pulley is operated by two cables

mixture strength is determined by the amount of fuel passed through the pilot jet, and also by the amount of air which is allowed to pass the pilot air screw. If the screw is turned in, this richens the mixture, and when the screw is turned out this weakens the mixture. The correct position for the air screws is normally one and a quarter turns out from the fully closed position.

7 The main carburation system comprises the main jet, the bleed pipe, the jet needle, the throttle valve, and the air jet. The main system comes into operation after the throttle is opened beyond one eighth of a turn of the throttle. It is only after this that sufficient vacuum is created at the jet needle to draw fuel up through the main jet. The fuel flows up through the main jet and bleed pipe, then between the needle jet and jet needle, and into the main bore where it is fully atomized. The fuel in fact is partially atomized before it reaches the bore, because the air bleed hole in the bleed pipe admits air to the fuel as it passes through the pipe.

8 When the throttle is opened, the slide rises up the bore of the carburettor. The jet needle is connected to this slide and because the needle is tapered, the more it is raised the more the fuel is allowed to flow. This is how engine speed increases. The cutaway on the slide regulates the air flow and the vacuum pressure into the carburettor bore. Finally, when the throttle slide is raised to its limit, the flow of fuel is dictated by the size of the main jet rather than the space between the jet needle and needle jet. The machine is then running on the main jet.

9 The float system is made up of the float, the float needle valve, and the valve seat. The fuel is maintained by the float assembly at a constant level in the float bowl, to meet the engines needs. As fuel flows into the bowl the float rises which in turn raises the float valve. When the float reaches a pre-determined height, the valve closes onto its seat and this shuts off the flow of fuel to the carburettor. Consequently as the engine uses the fuel, the level in the bowl drops, causing the valve to leave its seat and admit more fuel to flow into the float chamber. Cleanliness is the most important thing when working on the carburettors. To remove the carburettors loosen the four intake manifolds by undoing the crosshead screws in the clamps and remove the air cleaner hoses at the rear. Then pull the whole bank of four carburettors off.

10 To separate the carburettors, loosen the throttle cable mounting nuts, and disconnect the cables from the pulley. Remove the throttle stop screw locknut from both the carburettor to be removed and its companion, then detach the link piece. Remove the throttle stop screw and the screw spring, together with the spring seat. Remove the cap nut from the carburettor linkage of the carburettor that is to be removed, and then remove the spring and seat. **Note:** Be careful not to lose the spring that will rise up when the cap nut is removed.

11 Unscrew the four mounting screws from the mounting plate and remove the first pair of carburettors. It is easier to remove the carburettors in pairs as they are joined by a link. After all the carburettors have been removed from the mounting plate they are ready for dismantling.

5 Carburettors: dismantling and reassembly

1 The crossover lever and pulley, and the throttle return spring, need not be removed from the mounting plate, when dismantling the carburettors. The fuel may be drained from the float bowl by removing the drain plug and washer. Remove the top cover screws, then remove the cover and gasket, bend flat the locktab washer and unscrew the bolt from the operating arm. The operating arm can now be removed. Undo the two screws that secure the bracket assembly to the throttle slide, and lift the bracket complete with the operating arm and connector assemblies out of the carburettor bore.

2 Remove the throttle valve and the needle from the bore, taking care not to bend the needle. Remove the plunger assembly after first removing the lever, cap, and guide screw.

Undo the float bowl screws, remove the bowl and the gasket, then take out the hinge pin and remove the float and the float needle valve. Remove the main jet, the air bleed pipe. Invert the carburettor, and gently press out the needle jet with a wooden dowel. Remove the float valve seat, the pilot jet and the pilot air screw with spring.

3 Clean all the components in clean petrol and then blow them dry with compressed air, taking care to clear all passages. Inspect all the jets and the needle valve and seat, and renew them if they are worn, especially if there is a bright ridge round the needle valve and seat. It is best to renew these as a pair.

4 Inspect the float for leakage. Check whether petrol has entered the float by the weight of the float. If the float assembly is punctured it must be renewed.

5 Remove the main jet with a wide blade screwdriver, also inspect the needle jet for wear. After lengthy service the needle jet should be renewed along with the needle as these components are in continuous use. If not renewed petrol consumption will increase.

6 The carburettor slide should be able to slide down the carburettor bore by its own weight. If it will not do this even when lightly oiled, it will not function correctly.

7 Assembly of the carburettors is the reverse order of dismantling. Use new gaskets and 'O' rings. Do not overtighten the jets when installing into the carburettor body.

8 Make certain that the carburettor jet needle is replaced back in the same position as when it was removed. The needle clip should be in the third groove from the top.

6 Carburettors: adjustment

1 To check the float height adjustment with the carburettors in situ, first turn the fuel tap to 'OFF'. Then remove the carburettor vent tube. (Be prepared to catch the fuel that will run out.) Remove the float bowl drain plug. Install the Kawasaki fuel measuring devicce (Part number 57001-122) in place of the drain plug and hold the plastic tube against the carburettor body. Turn the fuel tap to the 'ON' position. The petrol level in the tube should be 0.10 - 0.18 inch (2.5 - 4.5 mm) below the edge of the carburettor body. If the petrol level is incorrect the float must be adjusted in the following manner. Drain the fuel from the float chamber and remove the chamber bowl. Be prepared to catch the float and float hinge pin, also the float needle. Bend the tang on the float slightly to adjust the float height. Bending the tang up will lower the fuel level. **Note:** When checking the fuel level of the two inside carburettors, the outside carburettor base may be used as a reference point for the measuring gauge.

2 Adjust the throttle cables by starting with the opening cable first. Loosen the locknut on the throttle opening cable, and use the adjuster to take up any slack in the cable before securing the locknut again. Loosen the locknut on the closing cable, and adjust it so there is about 1/16 inch (2 mm) of play in the throttle grip, then secure the locknut.

3 Perform the following tasks as a prelude to the actual adjustment of the carburettors at any time they are rebuilt or replaced, and especially if the engine idles roughly.

4 Remove the carburettors as described in Section 4 of this Chapter. Turn the throttle stop screw so that there is 3/8 inch (10 mm) between the bracket and the underside of the screw head. Loosen the closed throttle stopper locknut, and rotate the eccentric stopper screw until there is about 1/16 inch (1.5 - 2.0 mm) clearance between the stopper and the top of the pulley. The closed throttle stopper screw is set in the throttle cable pulley (see photo 4.2).

5 Locate the notch that is cut into the bottom of the throttle slide, then loosen the locknut and rotate the synchronising screw (see Fig. 2.4) until there is about 0.024 - 0.028 inch (0.6 - 0.7 mm) clearance between the notch and the bottom of the bore, then secure the locknut. **Note:** This is a very delicate operation and it must be carried out on each of the carburettors, so that the adjustment is **exactly the same** for all four.

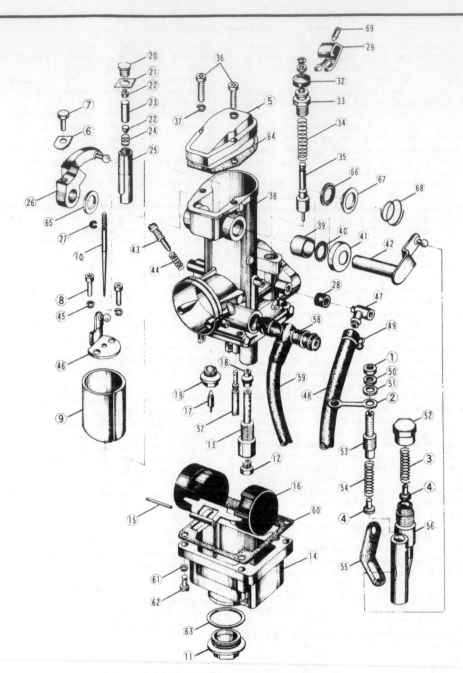

Fig. 2.4 Carburettor assembly

1 Synchronising screw locknut	18 Needle jet	35 Plunger assembly	52 Cap nut
2 Double washer link	19 Valve seat	36 Bolt	53 Synchronising scre
3 Spring	20 Guide screw	37 Lockwasher	54 Spring
4 Spring seat	21 Lock washer	38 Mixing chamber	55 Dust plate
5 Top cover	22 Spring seat	39 Hose	56 Connector
6 Lock washer	23 Pin	40 Washer	57 Pilot jet
7 Bolt	24 Spring	41 Cup	58 Fuel pipe fitting
8 Screw	25 Connector	42 Lever assembly	59 Fuel pipe
9 Throttle valve	26 Lever assembly	43 Air screw	60 Gasket
10 Jet needle	27 Circlip	44 Spring	61 Lock washer
11 Drain plug	28 Hose	45 Lockwasher	62 Bolt
12 Main jet	29 Lever	46 Bracket assembly	63 'O' ring
13 Air bleed pipe	30 Circlip	47 Air vent pipe fitting	64 Gasket
14 Float bowl	31 Ring	48 Air vent pipe	65 Spacer
15 Pin	32 Cap	50 Lock washer	66 Oil seal
16 Float	33 Guide screw	51 Washer	67 Collar
17 Float valve needle	34 Spring	52 Cap nut	68 Spring

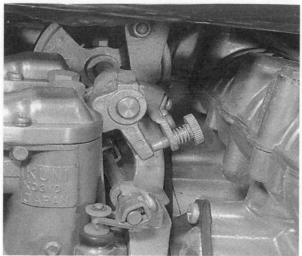

4.10 Throttle stop screw

5.1 Remove top cover screws and lift away cover

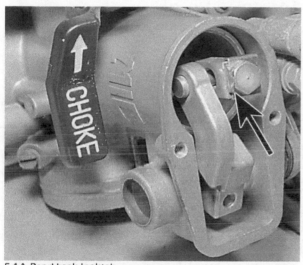

5.1A Bend back locktab

5.1B Remove seat

5.1C Pull out operating arm

5.2 Seat and holder in operating arm

5.2A Spring seat fits one way in arm

5.2B Removing carburettor float bowl

5.2C Withdraw float hinge pin

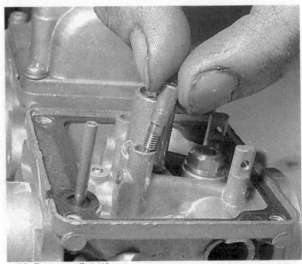

5.2D Remove pilot jet

5.2E Taking out jet holder and main jet

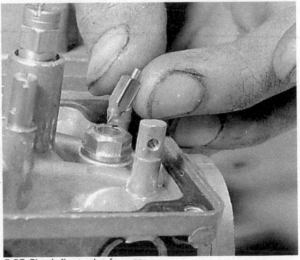

5.2F Check float valve for wear

6 Open the throttle by rotating the pulley until the bottom edge of the lowest of the four throttle valves is even with the top of the carburettor bore, then rotate the open throttle stop screw so that the pulley is stopped at this point.

7 Screw the pilot air screws into their seats gently and then turn each one out 1½ turns on all four carburettors, so that they are all equally adjusted.

8 To adjust the idling speed, start the engine and run for about five minutes until the normal operating temperature is reached, then adjust the engine idle speed using the throttle stop screw until the idle speed is about 800 - 1,000 rpm, registered on the tachometer. Adjust the pilot air screw on each carburettor to the position where the highest idle speed is reached. If the idle speed exceeds the limits given in the previous step, lower the speed to within the limits using the idling stop screw.

9 Turn in each air screw evenly only a fraction of a turn at a time, a quarter or half turn on each occasion, and then readjust the idle speed to 800 - 1,000 rpm.

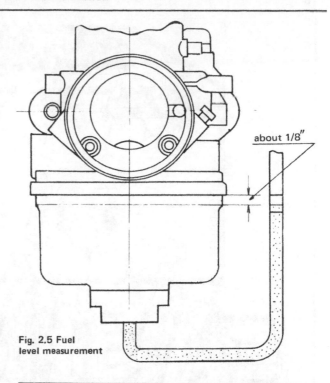

Fig. 2.5 Fuel level measurement

about 1/8"

5.8 Pilot air screw (arrowed)

6.1 Bend tang to alter float height

7 Carburettors: synchronising

1 Adjust the inlet manifold vacuum of each carburettor using the following method. This is a method that does not involve the use of vacuum gauges. The machine should be standing in a good airflow to keep the engine operating temperature down. Pay careful attention tho the exhaust note, and place your hands behind the silencers to feel the exhaust pressure. Compensate for any variation in exhaust pressure by making adjustments to the synchronising screw for the carburettor in question. A box spanner and a small screwdriver can be used to slacken the locknut and rotate the synchronising screw. The task is made easier by raising the petrol tank at the rear, and propping it up with a suitable stay.

2 Readjust the air screw on any of the carburettors that have to be readjusted, then readjust the idling speed to within 800 - 1,000 rpm, by adjusting the throttle stop screw.

3 To adjust the carburettors with vacuum gauges, first remove the rubber caps from the vacuum fittings on the cylinder head on the early models, or on the carburettor intake manifolds on the later models. The vacuum gauges can now be attached, one hose to each of the four pipes, so that the four cylinders can all be read on the corresponding gauges. With the engine running at idle speed, close down the vacuum gauge intake valve until the gauge needle flutters less than 2 cm h.g. (0.8 in. h.g.).

4 The normal manifold vacuum gauge reading is 20 - 23 cm h.g. (8.9 in. h.g.) for each cylinder. If any gauge reads less than 15 cm h.g. (6 in. h.g.) recheck the pilot air screw adjustment, also make sure that the carburettor hose clamps and spark plugs are secure.

5 Balance the carburettors by adjusting the synchronising screws (see Fig. 2.4). All the carburettors should be adjusted to within 2 cm h.g. (0.8 in. h.g.). of each other.

6 Open the throttle fairly rapidly and allow it to snap shut several times, while watching to see if the vacuum gauge readings remain the same. Readjust any carburettors whose readings have changed.

7 The vacuum gauges can now be removed. Replace the protective rubber covers on the adapters. Readjust any carburettor by the pilot screw and adjust the idle speed to about 800 - 1,000 rpm.

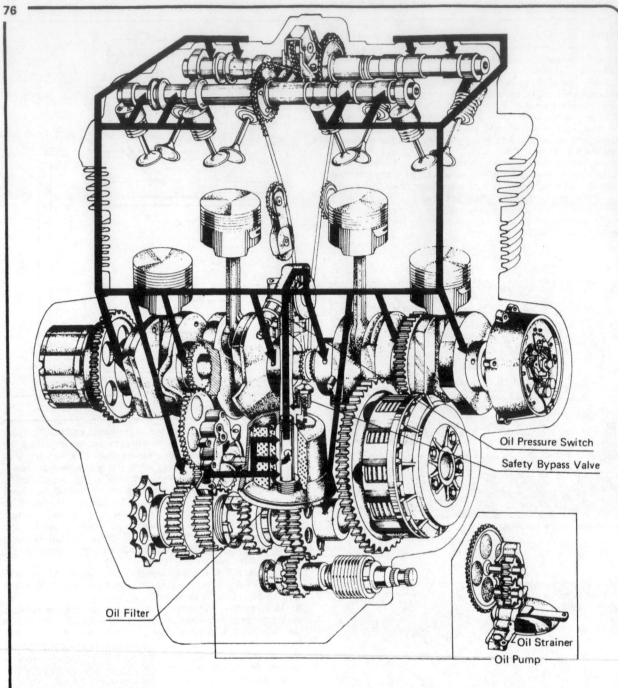

Oil Pressure Switch

Safety Bypass Valve

Oil Filter

Oil Strainer

Oil Pump

Fig. 2.6 The lubrication system

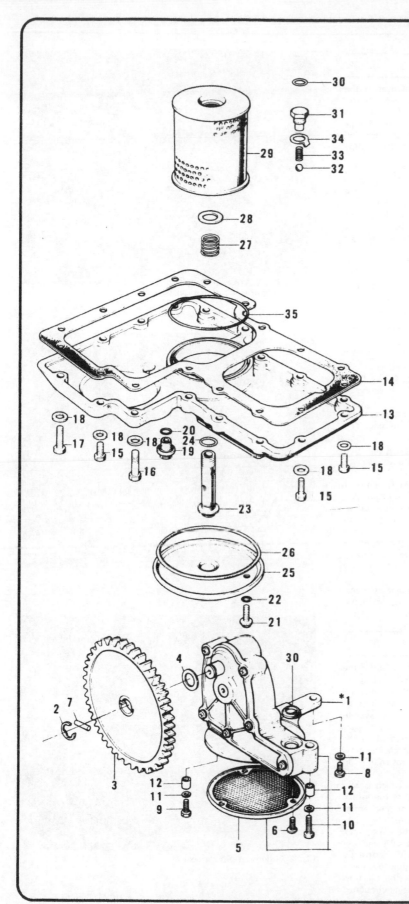

Fig. 2.7 Oil pump, oil filter and sump

1 Oil pump assembly
2 Circlip
3 Oil pump gear 49.7.
4 Oil pump washer
5 Oil pump gauze filter
6 Pan head screw - 3 off
7 Dowel pin
8 Bolt
9 Bolt
10 Bolt
11 Plain washer - 3 off
12 Dowel pin - 2 off
13 Oil sump cover
14 Oil sump cover gasket
15 Bolt for sump - 13 off
16 Bolt for sump - 3 off
17 Bolt for sump
18 Plain washer - 17 off
19 Oil drain plug for sump
20 'O' ring
21 Oil drain plug for filter
22 'O' ring
23 Oil filter bolt
24 'O' ring
25 Oil filter cover
26 'O' ring
27 Oil filter spring
28 Plain washer
29 Oil filter element
30 'O' ring - 2 off
31 Check valve bolt
32 Steel ball 3/8 inch
33 Check valve spring
34 Lock washer
35 'O' ring

8 Carburettors: settings

1 Some of the carburettor settings, such as the sizes of the needle jets, main jets, and needle positions are pre-determined by the manufacturer. Under normal riding conditions it is unlikely that these settings will require modification. If a change appears necessary, it is often because of an engine fault, or an alteration in the exhaust system eg; a leaky exhaust pipe connection or silencer.

2 As an approximate guide to the carburettor settings, the pilot jet controls the engine speed up to 1/8th throttle. The throttle slide cut-away controls the engine speed from 1/8th to 1/4 throttle and the position of the needle in the slide from 1/4 to 3/4 throttle. The size of the main jet is responsible for engine speed at the final phase of 3/4 to full throttle. These are only guide lines; there is no clearly defined demarkation line due to a certain amount of overlap that occurs.

3 Always err slightly towards a rich mixture as one that is too weak will cause the engine to overheat and burn the exhaust valves. Reference to Chapter 3 will show how the condition of the spark plugs can be interpreted with some experience as a reliable guide to carburettor mixture strength.

9 Air cleaner: dismantling, servicing and reassembly

1 The air cleaner is mounted immediately behind the four carburettors into which the carburettor intakes fit. The air filter housing contains the element that is removable for cleaning or replacement, when necessary.

2 To gain access to the air filter element, lift up the dualseat and remove the screen that covers the top of the housing. The element can now be pulled out from the top.

3 Clean the element with petrol or a cleaning solvent and then blow it dry with compressed air from the inside. Do not use any cleaner that will not completely evaporate.

4 Inspect the element and also the sponge gaskets for signs of wear or damage, and replace the element if either are damaged. The sponge gaskets can be glued back on if they are loose and in good condition. Be careful, when installing the element, not to crimp the gaskets.

5 The average useful life span of one of these elements is approximately 8,000 miles or 12 months, whichever the sooner, also if it has been cleaned three or four times due to use in very dusty conditions.

6 **Never** run the machine without the air cleaner element, otherwise the permanently weak mixture that results will cause severe engine damage.

10 Engine and gearbox lubrication

1 As previously described at the beginning of the Chapter the lubrication system is of the wet sump type, with the oil being forceably pumped from the sump to positions at the gearbox bearings, the main engine bearings, and the cam box bearings, all oil eventually draining back to the sump. The system incorporates a gear driven oil pump, an oil filter, a safety by-pass valve, and an oil pressure switch. Oil vapours created in the crankcase are vented through a breather to the air cleaner box, where they are recirculated to the crankcase, providing an oil-tight system.

2 The oil pump is a twin shaft, dual rotor unit which is gear driven off the crankshaft. An oil strainer is fitted to the intake side of the pump, which serves to protect the pump mechanism from impurities in the oil which might cause damage.

3 An oil filter unit is housed in the bottom of the sump, in an alloy canister containing a paper element. As the oil filter unit becomes clogged with impurities, its ability to function correctly is reduced, and if it becomes so clogged that it begins to impede the oil flow, a by-pass valve opens, and routes the oil flow around the filter. This results in unfiltered oil being circulated

9.2 Air cleaner is located under dualseat

11.2 Remove circlip retaining gear

11.2A Alignment pin for gear

throughout the engine, a condition which is avoided if the filter element is changed at regular intervals.

4   The oil pressure switch, which is situated at the top of the crankcase behind the cylinder block, serves to indicate when the oil pressure has dropped due to an oil pump malfunction, blockage in an oil passage, or a low oil content. The switch is not intended to be used as an indication of the correct oil level.

5   As previously mentioned an oil breather is incorporated into the system. It is mounted in the top of the crankcase, and is essential for an engine of this size with so many moving parts. It serves to minimise crankcase pressure variations due to piston and crankshaft movement, and also helps lower the oil temperature, by venting the crankcase. The breather tube carries the crankcase vapours to the air cleaner housing where they become mixed with the air drawn into the carburettors. If the breather hose or the ports inside the breather become blocked, pressure may build up to such a level in the crankcase that oil leaks will occur. If the oil level is too high in the sump, this may result in oil misting severe enough to cause the air cleaner to become oil saturated. This will lead to poor carburation. Avoid overfilling the sump.

6   Excessive oil consumption as indicated by a blue smoke emitting from the exhaust pipes, coupled with a poor performance and fouling of spark plugs, is caused by either an excessive oil buildup in oil breather chamber, or by oil getting past the piston rings. First check the oil breather chamber and air cleaner for oil build up. If this is the fault, check the passageway from the air/oil separator in the oil breather chamber to the lower half of the crankcase. Blockage here will prevent oil flowing back into the crankcase, resulting in oil build-up in the breather chamber and air cleaner tube.

7   Be sure to check the oil level in the sump before starting the engine. If the oil level is not seen between the two marks adjacent to the sight 'window' at the bottom of the clutch cover, replenish with the correct amount of oil (8 pints) of the correct viscosity.

11.5 Check gears for wear and end float

11 Engine and gearbox lubrication: the oil pump

1   The engine oil pump is driven by gear from the crankshaft. The pump works on the gear principal, with a pair of gears running in unison.

2   To dismantle the pump for inspection, the sump has to be removed, and the two screws that hold the pump body to the crankcase. The pump can then be lifted away. Secure the pump in a soft-jawed vice, or take precautions to avoid deforming the pump body. Remove the circlip retaining the main gear, the alignment pin, and the shim.

3   Remove the six crosshead screws that hold the two pump body halves together, then gently tap the two shafts alternately until the halves can be separated, without damaging the shafts.

4   Clean the pump components thoroughly in petrol or a suitable solvent, then blow them dry. Inspect all the parts for a worn or damaged condition, and replace them as necessary. New gaskets and 'O' rings should be fitted.

5   Assemble the two internal gears in one side of the pump body, and check the clearance between each gear and the pump body with a feeler gauge. The standard clearance is 0.0001 - 0.0014 in. (0.003 - 0.036 mm). The pump must be replaced if it is worn beyond its serviceable limit of 0.004 in. (0.10 mm). The strainer screen should be inspected for damage or a permanently clogged condition and replaced if necessary. This is available as a separate replacement. The pump body and gears are sold as a complete unit only.

6   Assembly and re-installation of the pump is in the reverse order of dismantling. Make sure that the jointing surface of both pump body halves is in perfect condition, and is absolutely clean, and always use a new gasket during reassembly. Make sure that the pump drive gear meshes correctly with the crankshaft drive gear when replacing the pump. Also, when replacing the sump, make sure the gasket and surfaces are clean and that the

11.5B Remove gears to clean out pump

sump pan screws are tightened in a diagonal sequence to avoid deforming the gasket.

12 Rear chain lubrication: the chain oil pump

Z1 model only

1   The rear chain oil pump is located in the cover that shrouds the gearbox main drive sprocket. The pump is shaft driven, driving from the gearbox layshaft end. It is of the plunger type and is controlled by engine revolutions, since the layshaft drives the main sprocket and the pump shaft simultaneously. When the pump lever is adjusted for minimum flow (0 on a scale from 0 - 5), the length of plunger travel is at a minimum, and vice versa.

2   The pump is gravity fed from an oil tank mounted underneath the dualseat, through a hose. A non return check valve is mounted on the output side of the pump, and the oil flows through the hollow pump shaft to the output shaft where it exits through a series of drilled holes near the drive sprocket. As the output shaft rotates, the oil is flung onto the rear drive chain.

3   To remove the chain oil pump, unscrew the two crosshead mounting screws, and remove the pump. Insert a mounting screw into the feed hose to stop oil leaking from the hose.

4   Inspect all parts for a worn or damaged condition. The 'O'

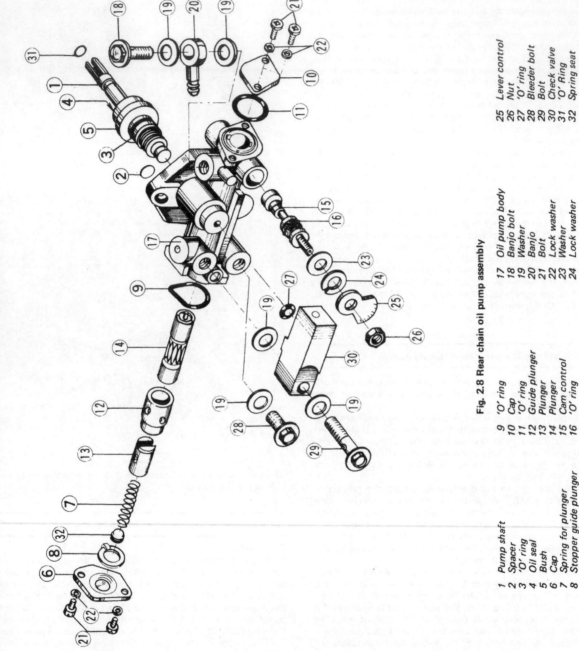

Fig. 2.8 Rear chain oil pump assembly

1	Pump shaft	9	'O' ring	17	Oil pump body	25	Lever control
2	Spacer	10	Cap	18	Banjo bolt	26	Nut
3	'O' ring	11	'O' ring	19	Washer	27	'O' ring
4	Oil seal	12	Guide plunger	20	Banjo	28	Bleeder bolt
5	Bush	13	Plunger	21	Bolt	29	Bolt
6	Cap	14	Plunger	22	Lock washer	30	Check valve
7	Spring for plunger	15	Cam control	23	Washer	31	'O' Ring
8	Stopper guide plunger	16	'O' ring	24	Lock washer	32	Spring seat

rings and the check valve are the only parts that normally wear, and consequently are the only available replacement parts If any other parts are worn or damaged, the complete pump assembly should be replaced.

5 Determine that the check valve will pass oil only in one direction, and that it is not clogged. The valve can be cleaned with petrol or solvent in a syringe. Whenever the pump is dismantled or if the oil supply tank runs dry while the pump is in operation, air will enter into the pump, which must then be bled. Do this by removing the bleeder bolt until oil starts to run from the hole, and then secure the bolt.

6 Assembly and installation are in the reverse order of removal. Lubricate the oil seal with clean oil before pressing it into position, and lubricate the 'O' and 'V' rings, plungers and plunger guide before assembly. Rotate the rear wheel until the pin in the layshaft is brought into line with the groove in the oil pump shaft. This is essential for correct installation as otherwise the pump will not seat correctly. Replace the mounting screws and secure them tightly. Bleed the pump by slackening the bleeder bolt until oil runs from the bolt and then secure the bolt. The pump cover can now be replaced. Check the level in the chain oil tank and replenish with SAE 30 or 40 oil. As previously mentioned the chain oil pump can be adjusted for oil flow 'O' supplies the least oil and 'S' is for maximum output. The chain should be kept just wet but not dripping. 'O' is usually the best position. The Z1-B models have dispensed with the chain oil pump altogether and use a prelubricated chain that requires attention only at infrequent intervals.

13 Fault diagnosis: Fuel system and lubrication

Symptom	Cause	Remedy
Engine will not start	Fuel starvation	Check if fuel flows to the carburettors when sucking main tube to fuel tap Check air vent in fuel tank cap. Check strainer and clean. Check float valves in carburettors. Clean and reset.
Engine runs erratically and will not pull	blocked jets	Remove jets, and clean.
Engine overheats and starts to pink	Shortage of oil	Check oiling system and refill if oil level is low
Engine starts to smoke excessively in traffic	Oil breathing system blocked	Clean out return oil passage from breather to sump
Rear chain throwing excessive oil	Chain oil pump set incorrectly (Z1 models only)	Remove oil pump cover and reset gauge to 'O'

Chapter 3 Ignition system

Contents

General description 1
Crankshaft alternator: checking the output 2
Ignition coils: checking 3
Contact breaker: adjustments 4
Contact breaker points: removal and replacement 5
Condensers: removal and replacement 6
Ignition timing: static setting 7
Ignition timing: stroboscopic setting 8
Automatic timing unit: removal, cleaning and
checking for wear 9
Spark plugs: checking and resetting the gaps 10
Fault diagnosis: ignition system 11

Specifications

Spark plugs

Make	NGK *
Size	14 mm
Type	B8ES (B9ES - high speeds)
	Alternative: Motorcraft AG1
Gap	0.028 inch to 0.031 inch (0.7 - 0.8 mm)
Contact breaker points gap	0.012 - 0.016 inch (0.3 - 0.4 mm)
Ignition timing	From 5° BTDC @ 1,500 rpm
	(20° BTDC @ 1,500 rpm - Z1-B)
	To 40° BTDC @ 3.000 rpm

Manufacturers recommendation

1 General description

1 The spark necessary to ignite the petrol vapour in the combustion chambers is supplied by a battery and two ignition coils (one coil to two cylinders).

There are two sets of contact breaker points, two condensers, four spark plugs and an automatic ignition advance mechanism. The breaker cam, which is incorporated in the advance mechanism, opens each set of points once in 180° of crankshaft rotation, causing a spark to occur in two of the cylinders. The other set of points fires 180° later, so that in every 360° of crankshaft rotation each plug is fired once. One extra spark occurs during the time when there is no combustionable material in the chamber.

Each set of points has one fixed and one movable contact, the latter of which pivots as the lobe of the cam separates them. The two condensers are wired in parallel, one with each set of contact points, and these function as electrical storage reservoirs, whilst also preventing arcing across the points. The condensers serve to absorb surplus current that tries to run back through the system when there is an overload situation, and feeds the current back to the ignition coils. They also help intensify the spark. When the points are closed, the current flows straight through them to earth. When they open, there is now an open circuit. If not for the condensers, the current may arc across the points causing them to burn and pit. When the condensers reach their capacity, they discharge the current back through the primary windings and eventually to the spark plug. Any time the points get badly burnt, it is advisable to renew them, and the condensers also.

Each of the two coils has two high voltage spark plug leads, and as in the case of points, one coil serves cylinders 1 and 4, and the other, cylinders 2 and 3.

The coils convert the low tension voltage into a high tension voltage sufficient to provide a spark strong enough to jump the spark plug air gap. If at any time a very weak or erratic spark occurs at the plug, and the rest of the ignition system is known to be in good condition, it is time to renew an ignition coil. Although coils normally have a long life they can sometimes be faulty, especially if the outer case has been damaged.

2 The automatic advance mechanism serves to advance the ignition timing as the engine R.P.M. rises. The mechanism is made up of two spring loaded weights which under the action of centrifugal force created by the rotation of the crankshaft fly apart and cause the contact points to open earlier. If the mechanism does not operate smoothly, the timing will not advance smoothly, or it may stick in one position. This will result in poor running in any but that one position. Sometimes the springs are prone to stretching, which can cause the timing to advance too soon. It is best to check the automatic advance mechanism, by carrying out a static timing test on the ignition followed by a strobe test. It is always best to check the motion of the weights by hand every 2000 miles and to clean and lubricate the unit at the same time.

3 The ignition system is operated by a key switch, mounted on a dash panel between the speedometer and the tachometer.

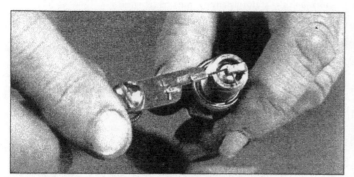

Electrode gap check - use a wire type gauge for best results

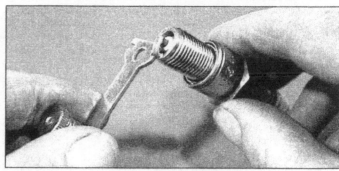

Electrode gap adjustment - bend the side electrode using the correct tool

Normal condition - A brown, tan or grey firing end indicates that the engine is in good condition and that the plug type is correct

Ash deposits - Light brown deposits encrusted on the electrodes and insulator, leading to misfire and hesitation. Caused by excessive amounts of oil in the combustion chamber or poor quality fuel/oil

Carbon fouling - Dry, black sooty deposits leading to misfire and weak spark. Caused by an over-rich fuel/air mixture, faulty choke operation or blocked air filter

Oil fouling - Wet oily deposits leading to misfire and weak spark. Caused by oil leakage past piston rings or valve guides (4-stroke engine), or excess lubricant (2-stroke engine)

Overheating - A blistered white insulator and glazed electrodes. Caused by ignition system fault, incorrect fuel, or cooling system fault

Worn plug - Worn electrodes will cause poor starting in damp or cold weather and will also waste fuel

There are three positions on the switch, OFF, ON, and PARK. In the OFF position all the circuits are turned off and the key can be removed from the switch. In the ON position the motorcycle can be started and all the lights and accessories can be used. The key cannot be removed from the switch when it is in this position.

In the PARK position, the tail light and parking light stays on, and the key can be removed from the switch. The charging of the battery that operates the ignition system is taken care of by an AC alternator that is mounted on the left-hand side of the crankshaft. This supplies current which is rectified by a rectifier, mounted on a panel alongside the voltage regulator, on the right-hand side of the machine below the dualseat.

2   Crankshaft alternator: checking the output

1   The alternator generates all the current required by the machines electrical circuits; the output is three phase alternating current (AC). The output is changed to direct current (DC) by the rectifier, the voltage being controlled by the voltage regulator. The alternator consists of a rotor and armature. Permanent magnets supply the magnetic field of the rotor, so that no slip rings or brushes are necessary. This makes the rotor practically maintenance free. The armature consists of three sets of coils wound on laminated steel cores. The coils are connected in a 'Y' pattern, so that there is always a smooth, ample supply of current available.

2   To check the output of the alternator, the battery and the rectifier must first be tested so that they are known to be good. If the battery shows less than the required 12 volts it should be fully charged.

3   Remove the right-hand side cover, and unplug the green regulator plug from the connector panel. Make sure that all the lights, indicators, are turned off then connect a voltmeter of 30 volts DC range, with the negative side of the voltmeter to the negative side of the battery and the positive side of the voltmeter to the positive side of the battery.

4   Start the engine, and run at 4,000 rpm. Note the meter reading. The voltage reading should be between 15 and 20 volts DC. A lower reading than this indicates the alternator is defective.

5   Turn off the engine and disconnect the voltmeter leads from the battery. Set an ammeter to the 12 amp DC range, unplug the wire that goes between the fuse and the starter relay, connect the positive meter lead to the white wire on the fuse side, and connect the negative meter lead to the white wire on the relay side. This places the ammeter in series with the rectifier and the battery so that the battery charging circuit current can be measured.

6   Note: If the ammeter is connected in series direct to the battery terminal instead of at the above point, do not use the electric starter to start the engine, otherwise the reverse starting current will damage the ammeter.

7   Turn on the ignition switch and start the engine. Hold the engine speed at 4,000 rpm and note the ammeter reading. The reading should be 9.5 amperes or more if it is normal. If the current is below this figure, the alternator is defective.

8   To determine if the trouble lies in the windings or the rotor, first turn off the engine and disconnect the blue plug from the connector panel. Using an ohmmeter with a scale of R.X.1, measure the resistance between each pair of the three alternator wires going to the connector plug: BLUE — PINK, BLUE — YELLOW, and PINK — YELLOW. The resistance between any two wires should be 0.45 to 6 ohms. Less than this resistance means that the coils are shorted out. A higher resistance or no reading at all means that the coils are open circuit. If the coils are found to be either shorted or open circuit, replace with a new stator assembly.

9   Using the highest resistance scale of the ohmmeter, measure the resistance between each alternator wire and earth (engine or frame). No reading is normal. Any meter reading indicates a

short, and the stator assembly must be replaced.

10   If the windings have a normal resistance, but voltage and current checks show the alternator to be defective, then the rotor magnets have probably weakened and lost their efficiency. The rotor must then be replaced with a new one.

3   Ignition coils: checking

1   The ignition coils are a sealed unit designed to give long life, and are mounted on the frame tubes in the upper cradle behind the steering head. The most accurate test of a ignition coil is with a three point coil and condenser tester (electrotester).

2   Connect the coil to the tester when the unit is switched on, and open out the adjusting screw on the tester to 7 mm (0:28 inch). The spark at this point should bridge the gap continously. If the spark starts to break down or is intermittent, the coil is faulty and should be renewed.

3   It is not practicable to effect a satisfactory repair to a faulty ignition coil.

2.3   Take off side panel

4.3   Adjustment screws for:

A   *Contact breaker gap*
B   *Ignition timing - individual cylinder pairs*
C   *Ignition timing - main backplate*

4.3A Slacken contact breaker screws (circled) and use screwdriver on tabs near upper screw to adjust contact breaker gap

9.3 Check plug caps for tracking

4 Contact breaker adjustments

1 To gain access to the contact breaker it is necessary to remove the cover plate screws and the cover on the right-hand front of the crankcase.

2 Rotate the engine by slowly turning it over with the kick-starter until one set of points is fully open. Examine the faces of the contacts for pitting and burning. If badly pitted or burnt they should be renewed as described in Section 5, of this Chapter.

3 Adjustment is carried out by slackening the screws on the base of the fixed contact, and adjusting the gap within the range 0.012 - 0.016 inch (0.3 to 0.4 mm) when the points are fully open by moving the base contact by means of the raised tabs each side of the upper screw. Use feeler gauges to set the gap, tighten the two screws and ro check once tight. Turn the engine until the other set of points is fully open, then check and adjust the contact gap. Once complete, check both contact breaker gaps again - the setting can sometimes change as the screws are tightenend.

4 Before replacing the cover and gasket, place a slight smear of grease on the cam and a few drops of oil on the felt pad. Do not over lubricate for fear of oil getting on the points, and causing poor electrical contact.

5 Contact breaker points: removal and replacement

1 If the contact points are badly burnt or worn, they should be renewed. Undo the two screws that hold the base of the fixed contact of each set of points, and remove the wire leading to the condenser, which will allow the points to be lifted off. Removal of the circlip on the end of the pivot pin will permit the moving contact point to be detach. Note the arrangement of the insulating washers.

2 Replace the contact points by reversing the order of removal, making sure the insulating washers are in their correct order before replacing the contact arms.

3 Check and if necessary readjust the contact breaker gap. When the points are fully open the correct gap is within the range 0.012 to 0.016 in (0.3 to 0.4 mm).

6 Condensers removal and replacement

1 There are two condensers contained in the ignition system each one wired in parallel with a set of points. If a fault

developes in a condenser, ignition failure is likely to occur.

2 If the engine proves difficult to start, or misfiring occurs, it is possible that a condenser is at fault. To check, separate the contact points by hand when the ignition is switched on. If a spark occurs across the points as they are separated by hand and they have a burnt or blackened appearance, the condenser connected to that set of points can be regarded as unserviceable.

3 Test the condenser on a coil and condenser tester unit or alternatively fit a new replacement. In view of the small cost involved it is preferable to fit a new condenser, and observe the effect on engine performance as a result of the substitution.

4 Check that the screws that hold the condensers to the contact breaker plate are tight, and also form a good earth.

7 Ignition timing: static setting

1 Remove the circular contact breaker inspection cover to gain access to the contact breaker assemblies and to the timing marks, which are viewed via the aperture at the top of the contact breaker baseplate. Before any attempt is made to adjust or check the timing make sure that the contact breaker gap settings are correct. (See Section 4).

2 Some means of identifying the exact point at which the contacts separate will be required. If a pocket multimeter is available this can be used set on the resistance scale. Alternatively, a small 12 volt bulb can be employed. Solder a probe lead to the bulb case, and a second lead to the base contact. If possible, fit small crocodile clips to the ends of the leads. Whichever method is employed, one of the probe leads should be attached to the contact breaker terminal or spring blade, and the remaining lead to a sound earth, such as a cylinder head fin.

3 Note that if a multimeter is used, the ignition should be left *off*, whilst with the bulb arrangement, it will be necessary to turn the ignition on. Assemble the test lamp or meter, attaching the probe lead to the left-hand contact breaker set (cylinders No. 1 and 4). Rotate the engine so that the 1 - 4 'F' mark is visible in the inspection aperture, ensuring that it aligns with the fixed alignment mark.

4 It will be seen that each contact breaker set is mounted on a pressed steel segment, which in turn is secured to the contact breaker baseplate. Slacken the two screws, which retain the left-hand segment, *just* enough to permit it to be moved. Move the plate slowly, watching the meter needle or bulb carefully. As soon as the points start to separate, the bulb will light or the meter needle will deflect. Tighten the screws, then rotate the engine a few times and check that the contact faces begin to

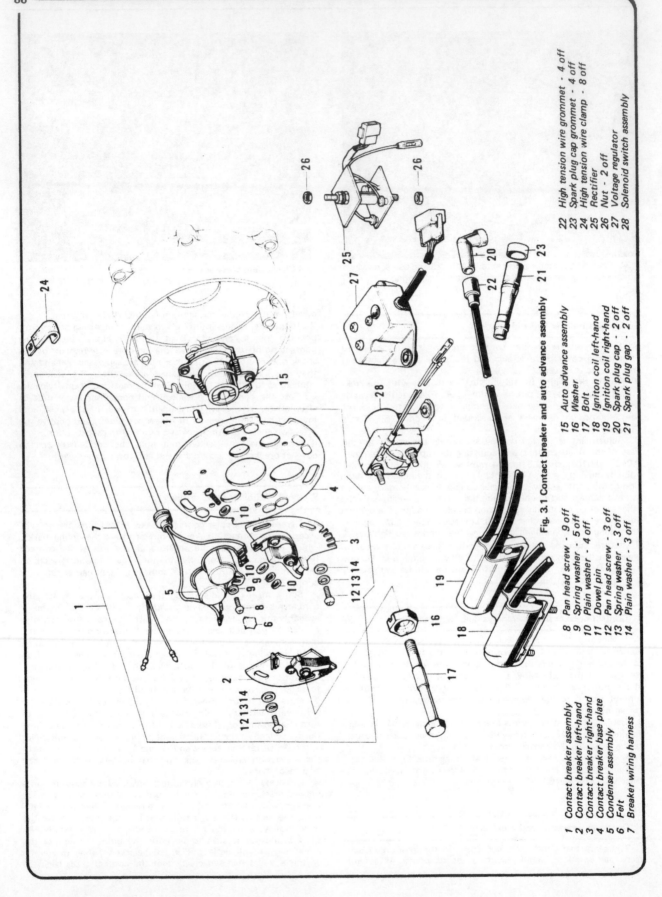

Fig. 3.1 Contact breaker and auto advance assembly

1 Contact breaker assembly
2 Contact breaker left-hand
3 Contact breaker right-hand
4 Contact breaker base plate
5 Condenser assembly
6 Felt
7 Breaker wiring harness

8 Pan head screw - 9 off
9 Spring washer - 5 off
10 Plain washer - 4 off
11 Dowel pin
12 Pan head screw - 3 off
13 Spring washer - 3 off
14 Plain washer - 3 off

15 Auto advance assembly
16 Washer
17 Bolt
18 Igniton coil left-hand
19 Ignition coil right-hand
20 Spark plug cap - 2 off
21 Spark plug gap - 2 off

22 High tension wire grommet - 4 off
23 Spark plug cap grommet - 4 off
24 High tension wire clamp - 8 off
25 Rectifier
26 Nut - 2 off
27 Voltage regulator
28 Solenoid switch assembly

separate just as the timing marks align. If all is well, repeat the procedure with the right-hand contact breaker assembly.

8 Ignition timing: stroboscopic setting

1 If a stroboscopic timing lamp is available, it can be used to check the timing quickly and accurately. If the timing has been lost completely as a result of dismantling it will be necessary to set the static timing first to enable the engine to be run. Two types of stroboscopic lamp, or 'strobes' are available. The cheaper neon types are best used under some form of shade, as the flash produced is not very intense. It may also be useful to paint white lines on the timing marks to make them more visible. If an xenon type strobe is to be used, connect it to a *separate* 12 volt battery to avoid spurious flashes from the machine's electrical system.

2 Connect the lamp up, following the manufacturer's instructions. Start the engine and set the throttle stop screw to give an idle speed of 800 - 1000 rpm. Aim the light at the aperture at the top of the contact breaker plate, and observe the timing marks which will appear frozen by the light pulses. If necessary, slacken the segment securing screws of the contact breaker assembly being dealt with, and move the segment until the timing marks coincide.

3 Repeat the operation for the remaining contact breaker set. The strobe may be used to assess the general condition of the ignition system during the timing check. If the image of the timing marks on the automatic timing unit appear to be blurred or if they jump about during the test, all is not well with the timing unit. Wear will have allowed the various parts of the unit to move slightly during running, and if badly worn it may be impossible to set the timing accurately.

4 Watch the timing mark, whilst slowly opening the throttle. The mark should appear to move as the timing unit advances, until at full advance, a second mark is in line with the fixed timing mark. Full advance should be reached by about 3000 rpm. If the timing fails to advance, or if the image is erratic, the timing unit should be removed, cleaned and examined for wear after releasing the contact breaker baseplate.

9 Automatic timing unit: removal, cleaning and checking for wear

1 After a long period of use, the action of the timing unit will tend to become erratic and sluggish, and this condition will be indicated during a stroboscopic timing check as described in Section 8 of this Chapter. To gain access to the unit, it is neces-
sary to release the contact breaker baseplate which is secured by three screws. Before removing the baseplate, mark its position in relation to the cylinder head casting. This will enable the correct timing to be closely approximated after installation, facilitating a stroboscopic timing check without first having to set the timing statically.

2 Check the action of the contact breaker cam by twisting it. The cam should move smoothly as the bob-weights move outwards. The unit can be removed after releasing the centre bolt which retains it to the camshaft end. If the unit is dismantled for cleaning, make a sketch first, noting the position of the cam lobe in relation to the base of the unit. This is important, as the cam may otherwise be refitted 180º out.

3 Clean the unit carefully in clean petrol or methylated spirit to remove any accumulation of dirt. Check the moving parts for wear after cleaning, rejecting the unit if it is sloppy in operation. Little can be done by way of reclamation, and a new unit should be fitted if it is worn. When reassembling the unit, grease the bob-weight pivot pins, and fill the groove in the centre spindle with grease before refitting the cam.

4 Refit the timing unit to the camshaft end, ensuring that the locating pin is aligned correctly. Do not forget to check the ignition timing after the contact breaker baseplate has been fitted.

10 Spark plugs: checking and resetting the gaps

1 The spark plugs fitted to the Z1 series are 14 mm NGK B8 ES this is the recommendation for standard road usage. Certain operating conditions may indicate a change in the grade of spark plug, although the type recommended by the manufacturer usually gives the best all round service and performance. For high performance use a grade higher - B9ES.

2 To reset the plug gap always gently tap the outer electrode. The special plug tool as shown in the diagram is the best method of making this adjustment without risk of damage. The correct gap is from 0.7 mm to 0.8 mm (0.028 — 0.031 inch).

3 When screwing the spark plugs into the cylinder heads, always use the plug spanner in the tool kit and do not over-tighten, otherwise damage will result to the threads of the cylinder head. If the plug threads are damaged they can usually be repaired by a competent dealer, using a Helicoil insert. These can be fitted at an economical price. Make sure the plug caps are a good fit, they should be kept clean to prevent tracking. The caps contain suppressors that eliminate both radio and TV interference

11 Fault diagnosis: ignition system

Symptom	Cause	Remedy
Engine will not start	Faulty ignition switch	Operate switch several times, in case the contacts are dirty. If the lights and other electrics function, the switch may need replacement.
Engine misfires	Faulty condenser	Replace with new condenser and retest.
	Faulty coil	Replace coil and retest.
	Fouled spark plugs	Replace with new plugs, and clean the originals.
	Poor spark due to generator failure, and discharged battery	Check the output of the generator. Remove and recharge the battery.
Engine lacks power and overheats	Retarded ignition timing	Check the contact gaps and ignition timing.
Engine fades when under load	Pre-ignition	Check the grade of plugs fitted. Change to recommended grade. Check ignition timing.

Chapter 4 Frame and forks

Contents

General description 1
Front forks: removal from the frame 2
Front forks: dismantling 3
Steering head bearings: examination and renovation ... 4
Front forks: examination and renewing 5
Front forks: replacement 6
Steering head lock 7
Frame examination and renovation 8
Swinging arm rear fork: dismantling, examination, and re-
novation 9

Rear suspension units: examination 10
Centre stand examination: 11
Prop stand examination: 12
Footrests and rear brake pedal: examination and renovation 13
Dualseat: removal and replacement 14
Speedometer and tachometer heads: location, examination 15
Speedometer and tachometer drives: location, examination 16
Cleaning the machine 17
Fault diagnosis: frame and forks: 18

1 General description

1 The frame of the Kawasaki Z1 series is of the full cradle type, in which the engine is supported by duplex tubes at the base of the crankcase.

2 A top tube runs from the steering head to a position at the rear of the petrol tank; the frame is extended to the rear mudguard with provision for fitting the mudguard. Lugs for the attachment of the dualseat, the pillion footrests, rear brake pedal, centre stand and prop stand are fitted to the frame.

3 The front forks are hydraulically damped, consisting of two telescopic shock absorber assemblies, each of which comprises an inner tube, an outer tube, a spring and a cylinder, piston and valve. The whole fork assembly is attached to the frame by the steering head stem and is mounted on two bearing assemblies contained in the steering head housing.

4 The damping action of the fork is accomplished by the flow resistance of the fork oil flowing between the inner and outer tubes. The method of removal, dismantling, and reassembly of the complete fork assembly is described in the following text.

2 Front forks removal from the frame

1 The only time the front forks need to be removed from the frame as a complete unit is for renewal of the steering head races or if the machine suffers frontal impact in an accident.

2 Commence operations by removing the handlebar levers with their control cables attached.

3 Detach the handlebars by undoing the four top bolts holding the top clamps. Take off the handlebars, leaving the bottom clamp halves integral with the top fork yoke.

4 Disconnect the speedometer and the tachometer drive cables from their respective instrument heads, pull out the bulb holders from the back of the instruments once the covers have been removed, and also from the instrument panel (the bulbs are a push fit). Remove the mounting bolts that secure the instruments and ignition cluster assembly and remove the whole assembly.

5 Unscrew the three headlamp fixing bolts, one each side and one underneath. Take the snap connectors apart on the headlamp harness, and detach the headlamp from the machine.

6 The machine should be stood firmly on its centre stand. Place a support, such as a very stout wooden box, or a small jack using a piece of wooden board as a bearer for the jack, underneath the crankcase, so that the front wheel is well clear of the ground. Remove the speedometer cable from the front wheel.

7 Remove the four nuts that secure the front wheel spindle clamps, and then remove the clamps. The front wheel can now be removed. The four bolts that secure the front mudguard can be undone and the mudguard removed. **Note:** Under no circumstances operate the front brake while the front wheel disc is removed, or the caliper piston will be forced out of the cylinder.

8 Remove the caliper by undoing the two mounting bolts and place it out of the way. The brake pipe need not be disconnected from the caliper, however the assembly should either be tied to the frame, or rested on a suitable receptable, to avoid bending the brake pipe.

3 Front forks dismantling

1 It is advisable to dismantle each fork leg separately using an identical procedure. There is less chance of mixing the parts if this approach is adopted. Commence by draining the fork legs of oil. There is a drain plug in each leg located near the bottom of the outer slider. A Phillips screwdriver is necessary for this operation. Note that each drain plug has a copper sealing washer. Pump the forks up and down several times to expel all the oil.

2 The Allen bolts can now be removed from the bottom of the fork legs (these are recessed in the bottom of the outer lower slider). There are two of them, one to each fork leg. Removal of these Allen bolts releases the stanchion assembly.

3 Unscrew the fork cap bolts (these are also the filler caps) on the top of each fork leg, and remove the centre cap nut from the steering stem and the upper fork yoke.

4 Slacken the two top clamp bolts in the upper yoke, slacken the two clamp bolts in the lower yoke, and remove the four bolts. Use flat bladed screwdrivers to spread the clamps in the yokes which will permit the fork leg to be withdrawn more easily.

5 The spring and stanchion assembly can now be pulled out. Remove the circlip from the stanchion using circlip pliers, and remove the stanchion from the lower fork leg.

6 Remove the dust seal from the lower fork leg, and then

remove the circlip and the oil seal. The fork leg is now completely dismantled. Repeat for the other leg.

4 Steering head bearings: examination and renovation

1 Before commencing reassembly of the forks, examine the steering head races. The ball bearing tracks of the cups and cones should be polished and free from indentures or cracks. If there are signs of damage or wear, the cups and cones must be renewed. New ball bearings should be fitted at the same time.
2 The cups are a tight push fit and should be drifted out of position.
3 There are thirty nine ¼ inch ball bearings fitted in the steering head, 19 in the top race and 20 in the bottom race. They should be packed with new grease to hold them into position.

5 Front forks: examination and renovation

1 Clean all the fork components in a suitable solvent, and then blow them dry. Inspect all the components for wear, stripped threads, scoring or other damage and renew where necessary.
2 Inspect the stanchion for straightness by rolling it on a flat surface. If it is not too badly bent it is possible for an expert repairer to straighten it with a lathe and press. If it is severely bent however, it must be replaced with a new one.
3 Inspect the springs for signs of compression. After lengthy service, the springs can often be renewed to advantage. Measure the spring to check its free length. If either spring has compressed (sagged) to less than 485 mm (19.1 in) on the Z1 model, or less than 496 mm (19.53 in) on Z1A or Z1B models, the springs in both fork legs must be renewed.

6 Front forks replacement

1 Replace the front forks by reversing either of the dismantling procedures described in Sections 2 and 3 of this Chapter, whichever is appropriate. Note that the damper unit must be locked into the fork by screwing in the Allen screw at the base, before the stanchion and spring are inserted.
2 Refit new oil seals to the lower fork legs and new circlips, if the old ones have opened too much.
3 Replace the fork legs in the bottom and top yokes. A flat bladed screwdriver will again be needed to prise open the fork yokes to facilitate entry of the stanchions.
4 Before tightening the top fork yoke make sure the flange surface of the fork leg cap is level with the top of the yoke.

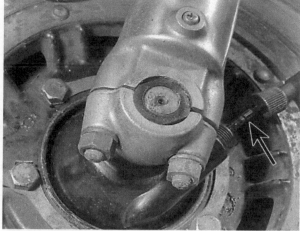
2.6 Remove speedometer cable from front hub

2.7 Removing the front wheel

2.7A Four bolts secure front mudguard

2.8 Removing caliper mounting bolts

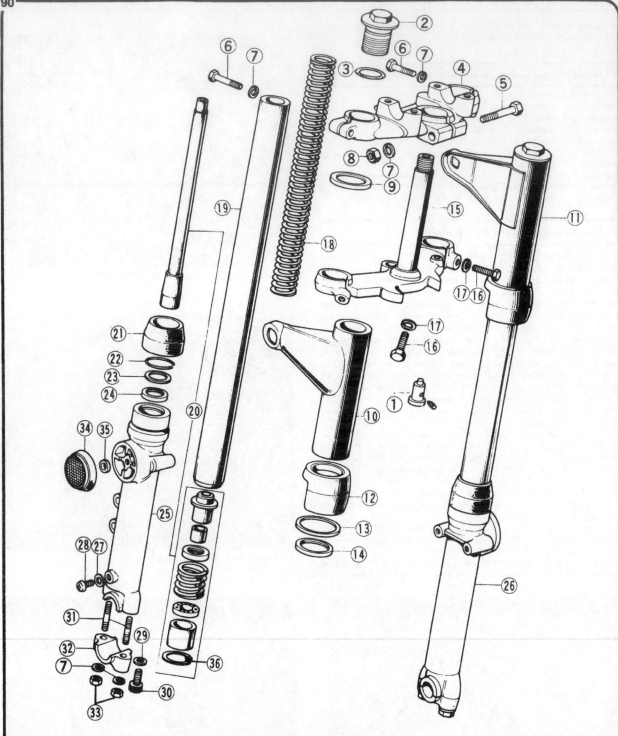

Fig. 4.1 Front fork components

1 Fork assembly complete	11 Right-hand cover	21 Fork dust shield - 2 off	30 Damper bolt - 2 off
2 Fork cap bolt - 2 off	12 Guide - 2 off	22 Circlip - 2 off	31 Stud - 4 off
3 'O' ring cap bolt - 2 off	13 Guide washer - 2 off	23 Plain washer - 2 off	32 Front spindle holder - 2 off
4 Steering stem top yoke	14 Gasket - 2 off	24 Oil seal - 2 off	33 Nut - 4 off
5 Bolt	15 Steering stem	25 Left-hand lower fork leg	34 Reflector - 2 off
6 Bolt - 2 off	16 Bolt - 2 off	26 Right-hand lower fork leg	35 Reflector rubber - 2 off
7 Spring washer - 7 off	17 Spring washer - 2 off	27 Fork drain plug gasket	36 Steering lock assembly
8 Nut	18 Fork spring - 2 off	- 2 off	37 Key set
9 Fork cover washer - 2 off	19 Stanchion - 2 off	28 Pan head screw - 2 off	38 Screw steering
10 Left-hand cover	20 Fork damper - 2 off	29 Damper bolt gasket - 2 off	lock

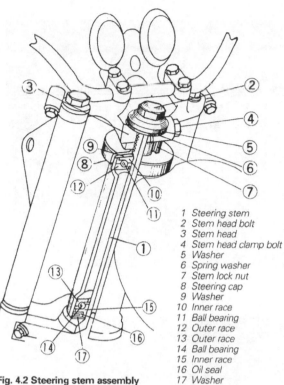

1 Steering stem
2 Stem head bolt
3 Stem head
4 Stem head clamp bolt
5 Washer
6 Spring washer
7 Stem lock nut
8 Steering cap
9 Washer
10 Inner race
11 Ball bearing
12 Outer race
13 Outer race
14 Ball bearing
15 Inner race
16 Oil seal
17 Washer

Fig. 4.2 Steering stem assembly

3.1 Removing the drain screw

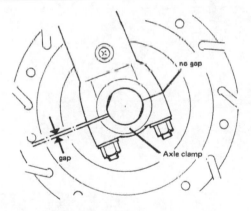

Fig. 4.3 Correct location of front spindle clamps

3.1A Allow oil to drain from leg

5 When the forks have been replaced, refit the front mudguard, by means of the four bolts inside the fork legs.
6 Refit the wheel in the bottom fork clamps, making sure they are correctly installed with the gap at the back, and no gap at the front.
7 Fill each fork leg with 169 cc of SAE10 fork oil. With the front wheel off the ground and the forks fully extended (ie not compressed), measure the distance from the top of the stanchion to the oil using a length of welding rod or a steel rule. On Z1 models the oil level should by 455 mm (17.9 in) and on Z1A and Z1B models it should by 475 mm (18.7 in). Add or remove oil as necessary so that the level is equal in each leg. Fit the fork cap bolts.
8 Before the machine is used on the road, check that the adjustment of the steering head bearings is correct. There should be no play at the steering head when the handlebars are pulled and pushed, with the front brake fully applied. The handlebars should swing from side to side with just a light tap, when the

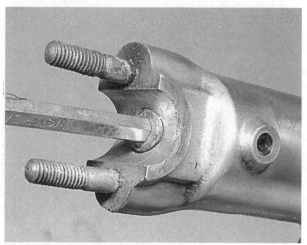

3.2 Use allen key for recessed bolt

3.3 Unscrew fork cap bolts

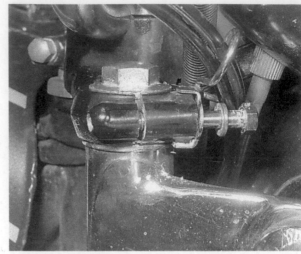

3.4 The top fork yoke bolts

3.4A The lower fork yoke bolts

3.4B Fork leg being withdrawn

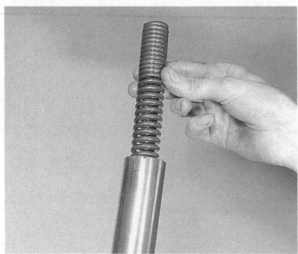

3.5 Take out spring

3.6 Removing stanchion

3.6A Take out stanchion with damper unit attached

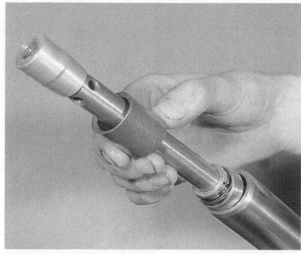

3.6B Check bushes for wear

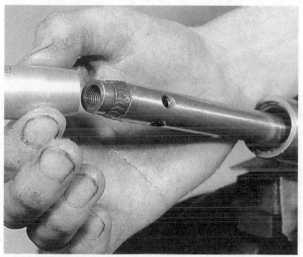

3.6C End bush on damper unit

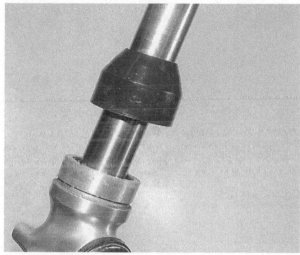

3.7 Remove dust seal

3.7A Take out circlip

3.7B Oil seal fits inside leg

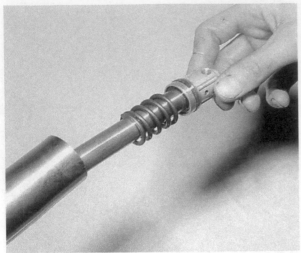

5.1 Clean all components before reassembly

5.1A Check bushes are a good fit

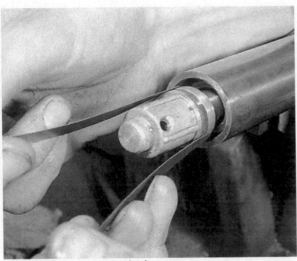

5.1 B Use feeler gauges to check wear

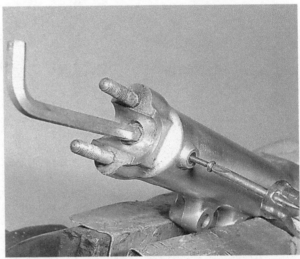

6.1 Lock screw with Loctite

6.4 Make sure flange surface is level with top of yoke

6.6 Refit wheel spindle

95

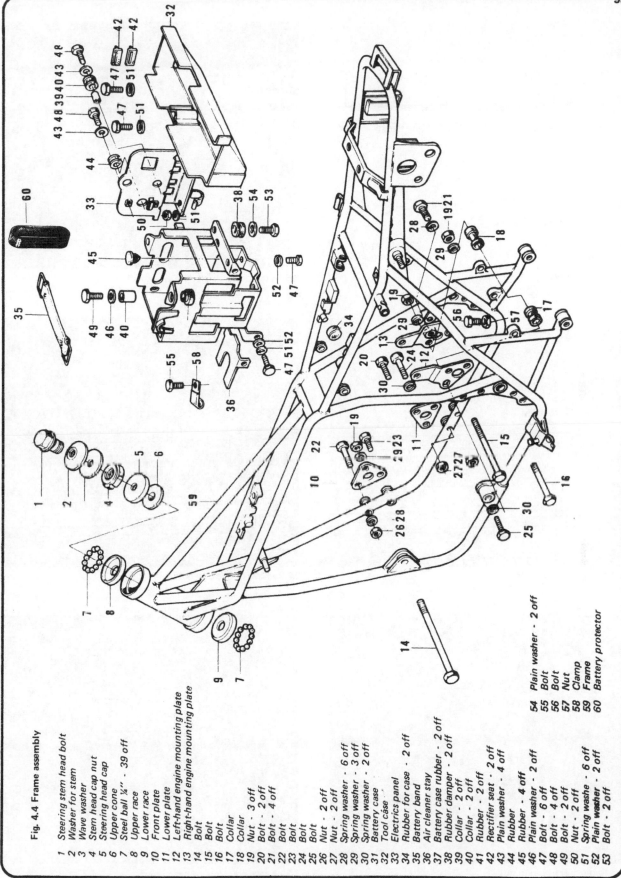

Fig. 4.4 Frame assembly

1 Steering stem head bolt
2 Washer for stem
3 Wave washer
4 Stem head cap nut
5 Steering head cap
6 Upper cone
7 Steel ball ¼" - 39 off
8 Upper race
9 Lower race
10 Front plate
11 Lower plate
12 Left-hand engine mounting plate
13 Right-hand engine mounting plate
14 Bolt
15 Bolt
16 Bolt
17 Collar
18 Collar
19 Nut - 3 off
20 Bolt - 2 off
21 Bolt - 4 off
22 Bolt
23 Bolt
24 Bolt
25 Bolt
26 Nut - 2 off
27 Nut - 2 off
28 Spring washer - 6 off
29 Spring washer - 3 off
30 Spring washer - 2 off
31 Battery case
32 Tool case
33 Electrics panel
34 Rubber for case - 2 off
35 Battery band
36 Air cleaner stay
37 Battery case rubber - 2 off
38 Rubber damper - 2 off
39 Collar - 2 off
40 Collar - 2 off
41 Rubber - 2 off
42 Rectifier seat - 2 off
43 Plain washer - 4 off
44 Rubber
45 Rubber - 4 off
46 Plain washer - 2 off
47 Bolt - 4 off
48 Bolt - 2 off
49 Nut - 2 off
50 Spring washe - 6 off
51 Plain washer - 2 off
52 Bolt - 2 off
53 Bolt - 2 off
54 Plain washer - 2 off
55 Bolt
56 Bolt
57 Nut
58 Clamp
59 Frame
60 Battery protector

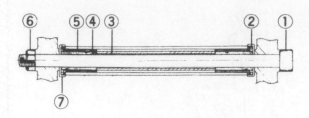

Fig. 4.5 Swinging arm pivot

1 Pivot shaft
2 'O' ring - 2 off
3 Collar
4 Sleeve
5 Bush - 2 off
6 Self-locking nut
7 Dust cap - 2 off

6.6A Clamp only fits one way round

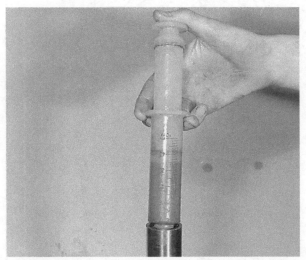

6.7 Using a plunger filler

9.2 Remove torque arm nut and bolt

9.3 Remove split pin in nut

9.3A Leave adjusters to hang down

machine is on the centre stand with the front wheel raised clear of the ground.

7 Steering head lock: removal and replacement

1 The steering head fork lock is located under the bottom fork yoke. It is of the barrel type with a tongue that extends from the barrel when the key is turned. To lock the steering, turn the handlebars to the left, insert the ignition key, and turn clockwise. If a fault developes in the lock it must be replaced with a new one. It is removed by undoing the single grub screw that screws in behind the fork yoke. UK models have a Neiman lock secured by a rivet.

8 Frame: examination and renovation

1 The frame is unlikely to require attention unless accident damage has occured. In some cases, renewal of the frame is the only satisfactory remedy if the frame is badly out of alignment. Only a few frame specialists have the jigs and mandrels necessary for resetting the frame to the required standard of accuracy, and even then there is no easy means of assessing to what extent the frame may have been overstressed.
2 After the machine has covered a considerable mileage, it is advisable to examine the frame closely for signs of cracking or splitting at the welded joints. Rust corrosion can also cause weakness at these joints. Minor damage can be repaired by welding or brazing, depending on the extent and nature of the damage.
3 Remember that a frame which is out of alignment will cause handling problems and may even promote "speed wobbles". If misalignment is suspected, as a result of an accident, it will be necessary to strip the machine completely so that the frame can be checked, and if necessary, renewed.

9 Swinging arm fork: dismantling and renovation

1 The rear fork of the frame is of the swinging arm type. It pivots on a shaft that passes through the crossmember and both sides of the main frame assembly, with a spacing collar, two inner bushes, and two flanged outer bushes. The whole assembly is held together with a pivot shaft that is bolted up with a nut one end.
   Worn swinging arm bushes can be detected by placing the machine on its centre stand and pulling and pushing vigorously on the rear wheel in a horizontal direction. Any play will be noticable by the leverage effect.
2 To remove the swinging arm fork, first position the machine on its centre stand, remove the rear brake adjuster and take out the torque arm bolt from the rear brake plate.
3 Remove the wheel by withdrawing the large split pin which locks the wheel nut, followed by the nut itself. Push the wheel forwards, having disengaged the adjusters, to allow the rear chain to be disengaged from the sprocket. Then pull the wheel rearwards out of the swinging arm. The chainguard can be removed by detaching the two fixing bolts.
4 Remove the lower two bolts that hold the suspension units to the swinging fork, so that the fork swings down. Leave the suspension units hanging from the frame, but slacken the top nut so that they are free to move. This facilitates reassembly.
5 Take out the swinging arm pivot shaft by undoing the nut on the left-hand side of the machine. This may need a gentle tap with a rawhide mallet and drift to displace it. Pull out the swinging fork complete with bushes.
6 Take off the two dust covers, noting the rubber seals ('O' rings) under the covers. This will expose the bushes. They should be drifted out from opposite ends along with the inner bushes and centre collar. Wash the bushes in a petrol/parrafin mix, then check the amount of play in them. If the clearance between the

bush and shaft exceeds 0.020 inch (0.5 mm) the bushes and shaft should be replaced as a set. The same applies if the pivot shaft is bent; it must be renewed.
7 Grease the pivot shaft and bushes prior to reassembly, and make sure the "O" rings are renewed if needed, also the dust covers. Always keep the pivot well greased. Apply a grease gun to the nipple provided until the smallest trace of grease is seen at the spindle end.

10 Rear suspension units: examination

1 Rear suspension units of the 5 way adjustable type with hydraulic damping are fitted to the Z1 series. The units can be adjusted to give 5 different settings. A hook spanner in the toolkit is used to adjust the units by means of peg holes.
2 There is no means of draining or topping up the units as they are permanently sealed. In the interests of good road holding, both units should be renewed if either starts to leak or loses its damping action.

11 Centre stand: examination

1 The centre stand is attached to the machine by two bolts on the bottom of the frame. It is returned by a centre spring. The bolts and spring should be checked for tightness and tension respectively. A weak spring can cause the centre stand to "ground" on corners and unseat the rider.

12 Prop stand: examination

1 The prop stand is secured to a plate on the frame with a bolt and nut, and is retracted by a tension spring. Make sure the bolt is tight and the spring is not overstretched, otherwise an accident can occur if the stand drops during cornering.

13 Footrests and rear brake pedal: examination

1 The footrests are of the swivel type and are retained by a clevis pin secured by a split pin. The advantage of this type of footrest is that if the machine should fall over the footrest will fold up instead of bending.
2 The rear brake pedal is held in position by a stud and domed nut the pedal return spring must be detached to remove the brake lever.

9.4 Remove bottom suspension unit bolts

9.5 Remove nut on pivot shaft

9.5A Tap out pivot shaft

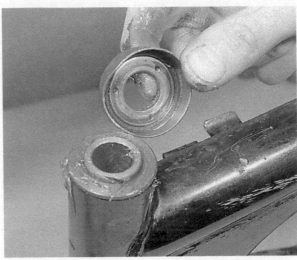

9.5B Pull down to release fork assembly

9.6 Take off dust caps

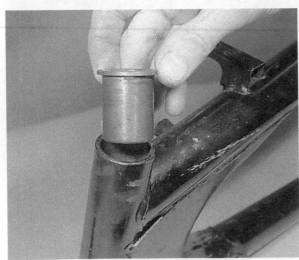
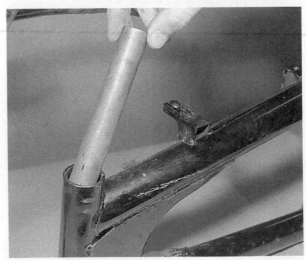

9.6A Remove end bushes

9.6B ... and distance centre collar

9.7 Grease bushes when assembling

9.7A Also pump grease through grease nipple

13.1 Make sure pillion rests are secure

15.1 Instruments are mounted on rubber bushes

16.1 Speedometer gearbox engages with tongues

14 Dualseat: removal and replacement

1   The dualseat is attached to the frame by two clevis pins that are located with split pins, on the right-hand side of the frame. To remove the seat, release the spring loaded catch on the left-hand side, and prop the seat up with the stay provided. Withdraw the two split pins from the clevis pivot pins, and remove the pivot pins. The seat mountings and damper rubbers can be left in place as the seat is lifted off.

2   If the dualseat is removed because it is torn, it is possible in most cases to find a specialized firm that recovers dualseats for an economical price, usually considerably cheaper than having to buy a new replacement. The usual charge is about 50% the cost of a new replacement, depending on the extent of the damage.

15 Speedometer and tachometer heads: removal and replacement

1   The speedometer and tachometer are both mounted together on a single panel on top of the front forks. They are mounted on studs with rubber bushes and secured with nuts. The heads are encased in light alloy shrouds secured to the instruments by a single crosshead screw. The shrouds have to be removed first, to enable the instruments to be released.

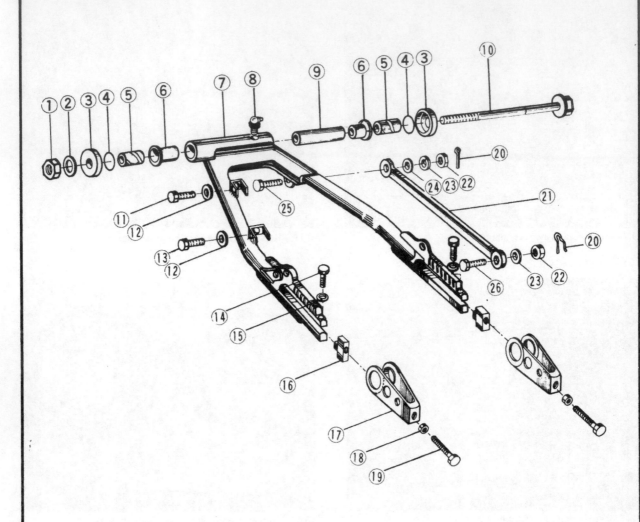

Fig. 4.6. Swinging arm fork

1	Self locking nut	10	Pivot shaft	19	Adjuster bolt
2	Washer	11	Chain guard mounting bolt	20	Cotter pin
3	Cap	12	Washer	21	Torque arm
4	'O' ring	13	Chain guard mounting bolt	22	Nut
5	Sleeve	14	Chain adjuster stop mounting bolt	23	Lock washer
6	Bush	15	Washer	24	Washer
7	Swinging arm	16	Chain adjuster stop mounting bolt	25	Bolt
8	Grease nipple	17	Chain adjuster	26	Bolt
9	Distance collar	18	Locknut		

2 After the shrouds are detached the drive cables can be un-screwed. The rubber mounted bulb holders can be pulled out with the bulbs. Check for blown bulbs while they are out. The four bulbs in the dash panel are also a push fit and can be checked at the same time.

3 The speedometer and tachometer heads cannot be repaired by the private owner, and if a defect occurs a new instrument has to be fitted. Remember that a speedometer in correct working order is required by law on a machine in the UK also may other countries.

4 Speedometer and tachometer cables are only supplied as a complete assembly. Make sure the cables are routed correctly through the clamps provided on the top fork yoke, brake branch pipe, and the frame.

16 Speedometer and tachometer drives: location and examination

1 The speedometer is driven from a gear inside the front wheel hub assembly. The gear is driven internally by a tongued washer (receiver). The receiver engages with two slots in the wheel hub, on the left-hand side. As the whole gearbox is pre-packed with grease on assembly, it should last the lift of the machine, or until new parts are fitted. The spiral pinion that drives off the internal gear is retained in the speedometer gearbox casing by a grub screw, which should always be secured tightly.

2 The tachometer drive runs off the camshaft in the cambox and screws directly into the cylinder head cover in the centre position. The cable is retained by a screwed ferrule, in the same manner as the speedometer cable.

17 Cleaning the machine

1 After removing all the surface dirt with warm water and a rag or sponge, use a cleaning compound such as "Gunk" or "Jizer" for the oily parts. Apply the cleaner with a brush when the parts are clog so that it has an opportunity to soak into the film of oil or grease. Finish off by washing down liberally, taking care that water does not enter into the carburettors, air cleaner or electrics. If desired, a polish such as Solvol Autosol can be applied to the alloy parts to give them a full lustre. Application of a wax polish to the cycle parts and a good chrome to the chrome parts will also give a good finish. Always wipe down the machine if used in the wet, and make sure the chain is well oiled. Check on models fitted with the rear chain oiler that the oiler is set correctly, also that there is always the required amount of oil in the chain supply oil tank. Check that the control cables are kept well oiled (this will only take 5 minutes of your time each week with an oil can). There is also less chance of water getting into the cables, if they are well lubricated.

18 Fault diagnosis: frame and forks

Symptom	Cause	Remedy
Machine veers to left or right with hands off handlebars	Wheels out of alignment. Forks twisted Frame bent	Check wheels and realign. Strip and repair. Strip and repair or renew.
Machine tends to roll at low speeds	Steering head bearings not adjusted correctly or worn	Check adjustment and renew the bearings if worn.
Machine tends to wander	Worn swinging arm bearings	Check and renew bearings. Check adjustment and renew.
Forks judder when front brake is applied.	Steering head bearings slack, front bushes worn	Strip forks and bushes; Check adjustment and renew.
Forks bottom	Short of oil	Replenish with correct viscosity oil.
Fork action stiff	Fork legs out of alignment. Bent shafts, or twisted yokes.	Strip and renew or slacken clamp bolts, front, wheel spindle and top bolts. Pump forks several times, and tighten from bottom upwards.
Machine tends to pitch badly	Defective rear suspension units, or ineffective fork damping	Check damping action. Check the grade and quantity of oil in the front forks.

Chapter 5 Wheels, brakes and tyres

Contents

General description 1
Front wheel: examination and renovation 2
Front wheel bearings: examination and replacement ... 3
Front wheel: reassembly and replacement 4
Rear wheel assembly: examination and renovation ... 5
Rear wheel bearings: examination and replacement ... 6
Rear brake assembly: examination, renovation and re-assembly 7
Rear sprocket assembly: examination, renovation, re-placement 8
Rear cush drive: examination and renovation 9

Rear brake assembly: adjusting 10
Final drive chain: examination and lubrication 11
Front brake assembly: examination and renovation ... 12
Front brake disc: examination and replacement 13
Front brake master cylinder: examination and renovation 14
Front brake disc caliper: examination and renovation ... 15
Bleeding the hydraulic brake system 16
Wheel balancing and alignment 17
Tyres: removal and replacement 18
Tyres: valves and dust caps 19
Fault diagnosis: wheels, brakes and tyres 20

Specifications

Brake fluid ... ...   DOT 3 or DOT 4

Tyres
Front: ... ...   325 x 19 4 P.R.
Rear: ... ...   400 x 18 4 P.R.

Tyre pressures: Solo riding
Front: 26 P.S.I. ... ...
Rear: 31 P.S.I. ... ...

Tyre pressures: Pillion rider
Front: 26 P.S.I.
Rear: 36 P.S.I.

Brakes:
Front: ... ...   Disc, 296 mm x 35 mm
Rear: ... ...   Internal expanding... 200 mm

1 General description

The Z1 series have a 19 inch front wheel and a 18 inch rear wheel. Front tyres are of the ribbed tread pattern; the rear tyres have a block tread pattern. All models employ steel rims in conjunction with cast aluminium hubs. The front brake is of the hydraulic disc type, the rear wheel is the internal expanding drum type.

2 Front wheel: examination and renovation

1   Place the machine on its centre stand so that the front wheel is clear of the ground. Spin the wheel by hand and check the rim for alignment.

Small irregulalities can be corrected by tightening the spokes in the affected area. Any flats in the wheel rim will be evident at the same time. In this latter case it will be necessary to have the wheel rebuilt with a new rim. The machine should not be run with a deformed wheel, since this will have a very adverse effect on handling.

2   Check for loose or broken spokes. Tapping the spokes is good guide to the correct tension; a loose spoke will always produce a different sound and should be tightened by turning the nipple in

3.2 Take out speedometer cable

3.2A Speedometer drive gear

3.2B Tongues engage with gearbox

3.2C Take out collar ....

3.2D ... and lift off wheel cap

5.2 Take out end bolts ....

5.2A ... withdraw wheel complete with spindle

5.3 Remove brake plate and shoes

5.3A Take out coupling ....

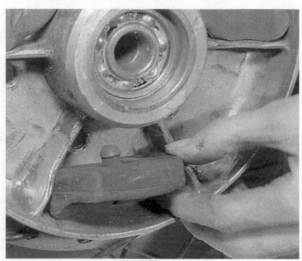
5.3B ...remove shock absorber rubbers

5.3C Use a ring spanner to slacken nuts

6.2 Remove wheel spindle

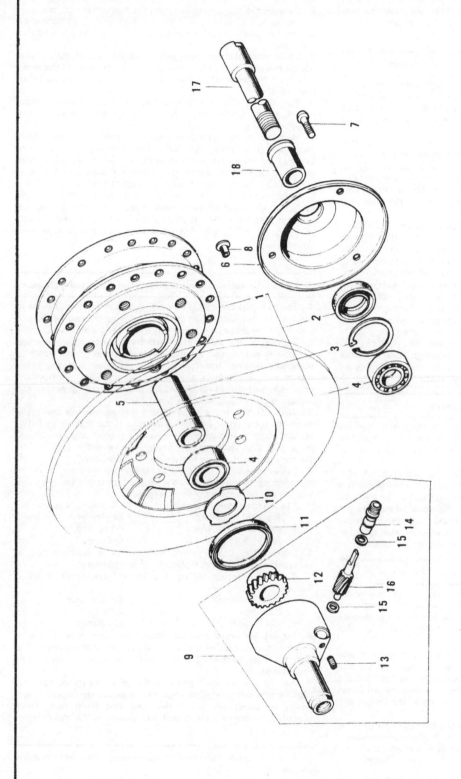

Fig. 5.1 Front hub components

1 Hub
2 Oil seal
3 Circlip
4 Ball bearing - 2 off
5 Bearing spacer
6 Dust cover

7 Counter-sunk screw - 3 off
8 Drum hole cap - 6 off
9 Speedometer gearbox
10 Oil seal
11 Speedometer gear reciever
12 Speedometer gear

13 Grub screw for bush
14 Speedometer cable bush
15 Thrust washer
16 Speedometer pinion
17 Front wheel spindle
18 Spindle collar

an anticlockwise direction. Always check run-out by spinning the wheel again. If the spokes have to be tightened an excessive amount it is advisable to remove the tyre and inner tube as detailed in Section 18 of this Chapter. This will allow the protruding ends of the spokes to be ground off, and prevent them giving rise to a spate of punctures.

3 Front wheel bearings: examination and replacement

1 Access is available to the front wheel bearings when the speedometer and front wheel spindle are removed. The bearings are of the ball journal type and non-adjustable. There are two bearings and two oil seals, the two bearings are interposed by a distance collar in the centre of the hub.
2 First remove the speedometer cable by undoing the knurled nut. To remove the bearings hold the speedometer gearbox stationary and unscrew the front wheel spindle, caution: do not hold the spindle and unscrew the gearbox otherwise the speedometer gear will be damaged. Take off the collar and wheel cap, and drive out the left-hand bearing using a double diameter drift from the right-hand side.
 When the bearing is removed, the distance collar can be taken out. Working from inside the hub, use the same drift to displace the right-hand bearing. Remove the oil seal, take out the retaining ring, and from the left side use the drift to tap evenly around the inner race of the right-hand bearing and knock it out.
3 Remove all the old grease from the hub and bearings, wash the bearings in petrol, and dry them thoroughly. Check the bearings for roughness by spinning them whilst holding the inner track with one hand and rotating the outer track with the other. If there is the slightest sign of roughness renew them.
4 Before driving the bearings back into the hub, pack the hub with new grease and also grease the bearings. Use the same double diameter drift to place them into position. Refit any oil seals or dust covers which have been displaced.

4 Front wheel: reassembly and replacement

1 Refit the speedometer gearbox, by holding the gearbox stationary and screw in the front wheel spindle. Do not hold the spindle and screw in the gearbox otherwise the speedometer drive gear will be damaged.
2 Have the bottom fork clamps ready when the front wheel is lifted back into position. First tighten the front spindle clamp bolt and then the rear bolt for each fork leg, so that there will be a gap at the rear after tightening. Spin the wheel to make sure it revolves freely, and check that the brake operates correctly. Turn the front wheel while inserting the speedometer cable, so that the tongue of the speedometer drive will locate correctly.

5 Rear wheel assembly: examination and renovation

1 Place the machine on the centre stand so that the rear wheel is raised clear of the ground. Check the rim for alignment, damage to the rim or broken spokes by following the procedure relating to the front wheel described in Section 2 of this Chapter.
2 Remove the bolt holding the rear torque arm to the brake plate. Disconnect the rear brake cable, and slacken and displace the rear wheel adjusters. Remove the split pin and nut from the wheel spindle and push the wheel forward so that the rear chain can be disengaged. Finally, ease the wheel back and free of the swinging arm.
3 Remove the wheel spindle and take out the brake plate. Take the coupling assembly from the cush drive rubbers, and remove the rubbers. The rear wheel sprocket can be unbolted for inspection, by removing the six nuts and the three locktabs.

6 Rear wheel bearings: examination and replacement

1 The rear wheel bearings are a drive fit into the hub. They are

separated by a spacer and a distance collar. There are two bearings in the hub and one in the centre of the rear final drive sprocket.
2 Remove the distance collar from the cush drive hub, take out the wheel spindle collar, and then the oil seal. Tapping evenly around the inner race from the inside of the coupling, knock out the bearing.
3 To remove the two bearings from the wheel hub, use a double diameter drive again and tapping evenly around the inner race from the sprocket side knock out the bearing on the brake plate side.
4 Remove the large distance collar, and tapping on the inner race from the brake plate side, knock out the bearing on the sprocket side.
5 Remove all the old grease from the bearings and hub. Wash the bearings in petrol and dry them thoroughly. Check the bearings for roughness by spinning them whilst holding the inner track with one hand, and rotating the outer track with the other hand. If there is the slightest sign of roughness renew them.
6 Before driving the bearings back into the hub and sprocket centre, pack the hub with new grease and also grease the bearings. Use the same double diameter drift to place them into position. Refit any oil seals or dust covers which have been displaced.

7 Rear brake assembly: examination, renovation and reassembly

1 The rear brake is of the internal expanding variety. Access to the brake shoes is obtained by first removing the rear wheel, and taking off the brake plate to which the shoes are attached.
2 To remove the shoes, first withdraw the two split pins and the double washer that secures the brake shoes. Use a punch to mark the original position of the brake cam and the brake operating lever. Remove the pinch bolt and lever. Remove the dust seal, and the brake shoes, by prying them up evenly and removing them along with the brake cam.
3 Take off the brake shoe return springs. Inspect the brake drum for a scored or warped condition. If the drum is scored or warped slightly, it is possible to have it turned down on a lathe by a specialist repairer but if the scoring is too deep or the warpage too great, a new replacement is necessary.
4 Inspect the brake shoes for excessive uneven wear, or for oil or grease on the linings. If the impregnation is too bad the shoes will have to be replaced with new ones.
 The standard measurement for the brake linings is as follows:

Standard thickness	Service limit
0.1909 - 0.2146 inch	0.118 inch
(4.85 - 5.45 mm)	(3.00 mm)

5 Inspect the brake return springs for a worn, pitted or collapsed condition, and replace them as necessary.
 The brake return spring free length specifications are as follows:

Standard length	Service limit
2.62 inch	2.72 inch
(66.5 mm)	(69.0 mm)

6 Inspect the brake cam and brake plate for signs of wear or damage, and replace as necessary. It cannot be overstressed that wear on these parts are critical if full braking efficientcy is to be maintained.
7 Assembly is in the reverse order of dismantling, use new locking tabs and split pins whenever possible, also smear a light touch of grease on the brake cam and pivot pins during assembly, taking care not to get any grease on the brake linings.

8 Rear sprocket assembly: examination, renovation and replacement

1 The rear wheel sprocket is held to the wheel by six nuts and three locktabs. To remove the sprocket, bend back the locktabs and undo the nuts. The sprocket needs to be renewed only if the

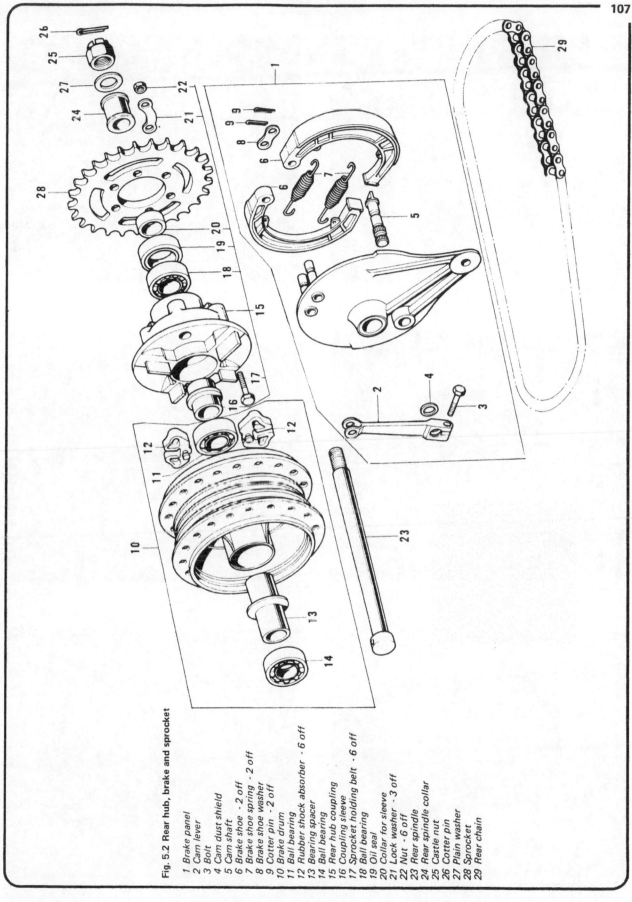

Fig. 5.2 Rear hub, brake and sprocket

1 Brake panel
2 Cam lever
3 Bolt
4 Cam dust shield
5 Cam shaft
6 Brake shoe - 2 off
7 Brake shoe spring - 2 off
8 Brake shoe washer
9 Cotter pin - 2 off
10 Brake drum
11 Ball bearing
12 Rubber shock absorber - 6 off
13 Bearing spacer
14 Ball bearing
15 Rear hub coupling
16 Coupling sleeve
17 Sprocket holding belt - 6 off
18 Ball bearing
19 Oil seal
20 Collar for sleeve
21 Lock washer - 3 off
22 Nut - 6 off
23 Rear spindle
24 Rear spindle collar
25 Castle nut
26 Cotter pin
27 Plain washer
28 Sprocket
29 Rear chain

6.2A Take out distance collar

6.2B Bearing in coupling, sprocket side

6.3 Bearing in wheel hub, shock absorber side

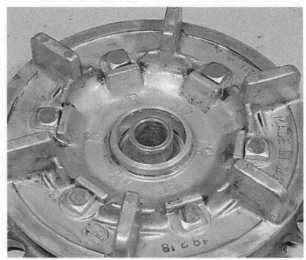

6.3A Bearing removed

7.2 Split pins locate shoes

8.1 The worn sprocket before removal

8.1A Note hooked teeth ....

8.1B ... compared with new sprocket

8.1C Fitting new sprocket

8.1D Bolts retain sprocket; tighten fully

teeth are worn, hooked or chipped. It is always good policy to change both sprockets at the same time, also the chain, otherwise very rapid wear will develop.

2   It is not advisable to alter the rear wheel sprocket size or the gearbox sprocket size. The ratios selected by the manufacturer are the ones that give optium performance with the existing engine power output.

9   Rear cush drive: examination and renovation

1   The cush drive assembly is contained in the left-hand side of the wheel hub. It takes the form of six triangular rubber pads incorporating slots. These engage with vanes on the coupling which is bolted to the rear sprocket. The rubbers engage with ribs on the hub and the whole assembly forms a shock absorber which permits the sprocket to move within certain limits. This cushions any surge or roughness in the transmission which would otherwise convey an impression of harshness.

2   The usual sign that shock absorber rubbers are worn is excessive movement in the sprocket, or rubber dust appearing in between the sprocket and hub. The rubbers should then be taken out and renewed.

10   Rear brake assembly: adjusting

1   If the adjustment of the rear brake is correct, the brake pedal will have a travel of 20 mm to 30 mm (0.8 to 1.2 inch). Adjustment is carried out at the end of the operating rod by the knurled adjuster.

2   It may be necessary to change the height of the stop lamp switch if the pedal travel has been altered to any marked extent. Raise the switch for the stop lamp to operate earlier by turning the adjustment nut clockwise.

11   Final drive chain: examination and lubrication

1   As the final drive chain is fully exposed on all models it requires lubrication and adjustment at regular intervals. To adjust the chain, take out the split pin from the rear wheel spindle and slacken the spindle nut. Undo the torque arm bolt, and leave the bolt in position, slacken the chain adjuster locknuts and turn the adjusters inwards to tighten the chain, or outwards to slacken the chain.

2   Chain tension is correct if there is (25 to 40 mm) about 1¼ inch slack measured at the centre of the bottom run of the chain between the two sprockets.

10.1 Range of lever travel is indicated

legs, squeeze the brake lever two or three times to confirm its operation, and bleed the air if necessary. (Refer to Section 16 of this Chapter, if it is necessary to bleed the brake system).

3 Do not run the chain too tight to try to compensate for wear, or it will absorb a surprising amount of engine power. Also it can damage the gearbox and rear wheel bearings.
4 Machines other than those fitted with a chain lubricating pump are equipped with a 'self-lubricating' chain. The latter should be cleaned and lubricated in situ in the same manner as conventional chains. As all chains are of the endless variety, removal for thorough cleaning involves a considerable amount of dismantling work. The manufacturers stress that the endless chain should not be broken or replaced with the conventional spring link type. For details of chain removal and maintenance, reference should be made to Chapter 7, Section 2.

12 Front brake assembly: examination and renovation

1 The hydraulic disc brake adopted for the Z1 series comprises four main components: A brake disc mounted on the front wheel hub, a master cylinder for pressurising the system, a brake hose for conveying the fluid pressure, and a caliper assembly, which presses the pads on to the brake disc by means of hydraulic pressure.
2 The brake disc attached to the front wheel rarely requires any attention. Renewal is necessary only if the surface is scored or damaged.
3 The disc is attached to the hub with six bolts and three lock-plates. If the front wheel is removed from the forks, the disc will be freed when the bolts are withdrawn. There is no necessity to detach the hydraulic brake system or to dismantle any part of it. The disc will pull clear of the friction pads in the caliper. It is advisable however, to place a clean spacer such as a piece of wood or metal between the pads, to prevent them being ejected if the front brake is unintentionally applied. This precaution is not necessary if the caliper assembly has also to receive attention.
4 Check the condition of the friction pads in the caliper. If either of the pads is worn up to the red line limit marked on the circumference, renew them as a pair by unscrewing the pad fastening screw from pad B. This is the single screw at the bottom. Take out the pad. The other pad is not secured by a screw and is ejected by squeezing the front brake several times. The pad will come out with fluid pressure when filling new pads. Apply brake pad grease (which forms part of the brake pad set) onto the pheriphery and backplate of the unretained pad in a very thin layer. Push the pad back into the caliper body, and then mount the other pad in the caliper body, screwing in from the inside of the fork leg. Install the front wheel into the fork

13 Front brake disc: examination and replacement

1 It is necessary to remove the front wheel, to detach the disc from the front wheel. Bend down the three lock tabs and remove the six disc mounting bolts, the disc can then be lifted off.
2 Clean off any oil on the surface of the disc using a degreasing agent. Inspect the disc for deep score marks or wear and measure the disc with a micrometer at its most work part. The disc can be safely skimed up to 0.050 inch to remove any score marks, but if the scoring or wear runs deeper than the disc's servicable thickness of 0.217 inch (5.5 mm) the disc must be renewed. The standard thickness for the disc is 0.276 inch (7.0 mm).
3 Check the disc for a warped condition. Warping will cause both the pads and disc plate to wear down quickly, and will also cause overheating and poor braking efficiency. Check the disc for warpage in the following ways: If the disc is still mounted on the wheel, the runout can be checked with a dial indicator. Alternatively, place the disc on a perfectly flat surface, and use a feeler gauge under the rim of the disc to check for clearance, which will indicate warpage. If the disc is warped in excess of 0.012 inch (0.3 mm) it must be renewed.

14 Front brake master cylinder: examination and renovation

1 The master cylinder and hydraulic reservoir take the form of a combined unit mounted on the right-hand side of the machine, on the handlebars, to which the front brake lever is attached. The master cylinder is actuated by the front brake lever and applies hydraulic pressure through the system to operate the front brake. The master cylinder pressurises the hydraulic fluid into the brake pipe, which being incompressible, causes the piston to move in the caliper unit and apply force to the friction pads on the disc plate. If the master cylinder seals leak, hydraulic pressure will be lost, and the braking action rendered much less effective.
2 Before the master cylinder can be removed for examination the system must be drained of fluid. Place a clean container below the caliper unit and attach a plastic tube from the bleed screw on top of the caliper unit to the container. Open the bleed screw one turn, and drain the system by operating the front brake lever until the master cylinder reservoir is empty. Close the bleed screw and remove the pipe.
3 Remove the banjo bolt that secures the brake line to the master cylinder, and then remove the rest of the brake line. Remove the clamp bolts which secure the master cylinder, and remove the cylinder. Remove the reservoir cap, the diaphragm, and then empty out any residual brake fluid. Remove the brake lever, and then the dust cover stopper, then remove the dust cover. Remove the retaining ring, using retaining ring pliers, then remove the stopper, the piston, the primary cup, and spring with the check valve from the master cylinder body.
  Caution: Do not attempt to remove the secondary cup from the piston as this will damage the cup.
4 The pressure switch, the three way fitting, the brake hose and brake line to the caliper, may now be removed.
5 Clean all the parts in a suitable solvent, then blow them dry. Use compressed air to blow out all the passages. A cloggged relief port will result in the friction pads dragging on the disc.
6 Inspect the master cylinder bore and pistons for signs of wear, rust, pitting or damage, and renew as necessary. The master cylinder and piston must also be renewed if worn past their servicable limit.

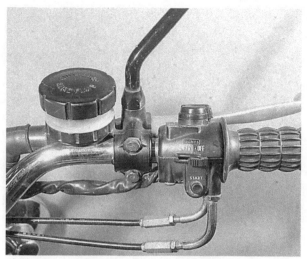

14.1 Master cylinder is mounted with two bolts

14.1A Resevoir cap removed

14.2 Undoing the bleed screw

14.4 Threeway branch connector

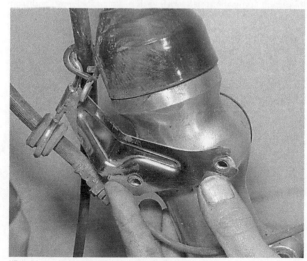

15.1 Mounting plate for brake pipe

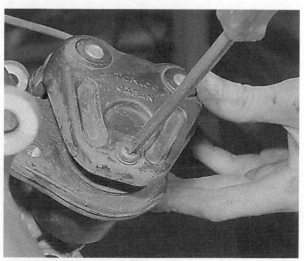

15.1A Remove securing screw

15.1B To release pad 'B'

15.2 Pad 'A' being removed

15.2A Shim in Pad 'A'

18.1 Tyre and load data is stamped on mudguard

7 Inspect the primary and secondary cups for signs of wear, damage, rotting or swelling and replace them as necessary. Leaking at the brake lever is an indication of bad cups. The piston must also be replaced if the secondary cup is damaged. It is preferable to replace the whole set of cups, piston, and spring, by means of the repair kit that is sold by the manufacturer. Replace the rubber dust cover if it is damaged or aged.
Inspect the fittings, hoses and brake pipes for signs of wear, rust, or cracking or any other damage and replace where necessary.

8 When assembling the master cylinder follow the removal procedure in reverse order. Pay particular attention to the following points: Make sure the primary cup is fitted the correct way round. Renew the lockwasher on the pivot bolt nut and fit it securely.

9 Mount the master cylinder on the handlebar so that the bottom clamp projection points towards the twist grip. Fully tighten the lower clamp bolt first and then the upper clamp bolt. Fill the reservoir with fresh brake fluid and bleed the system of air (air always accumulates in the system when new parts are fitted). Refer to Section 16 of this Chapter for the correct procedure. After the bleeding operation check the level of the reservoir and top up to the level line, if necessary.

10 The components parts of the master cylinder assembly and

the caliper assembly may wear or deteriorate in function over a long period of use. It is however generally difficult to forsee how long each component will work with proper efficiency, and from a safety point of view it is best to change all the expendable parts every two years on a machine that has covered normal mileage.

11 Note that hydraulic brake fluid is a very effective paint stripper. Avoid spillage, particularly on the petrol tank.

15 Front brake disc caliper: examination and renovation

1 When the system is drained of fluid, unscrew the brake pipe nut from the caliper body. Loosen the heads of the two allen shafts that hold the two halves of the caliper together, remove the two mounting bolts, and then take off the caliper assembly. Unscrew the Allen head shafts evenly, alternating between the two shafts a little at a time, then remove the right-hand inner caliper. Remove the right-hand friction pad by undoing the securing screw.

2 Remove the bolts which secure the caliper holder, taking care not to damage the boots, or "O" rings and then remove the outer or left-hand friction pad. Now remove the two shafts from the outer caliper then remove the piston dust seal, and pull the

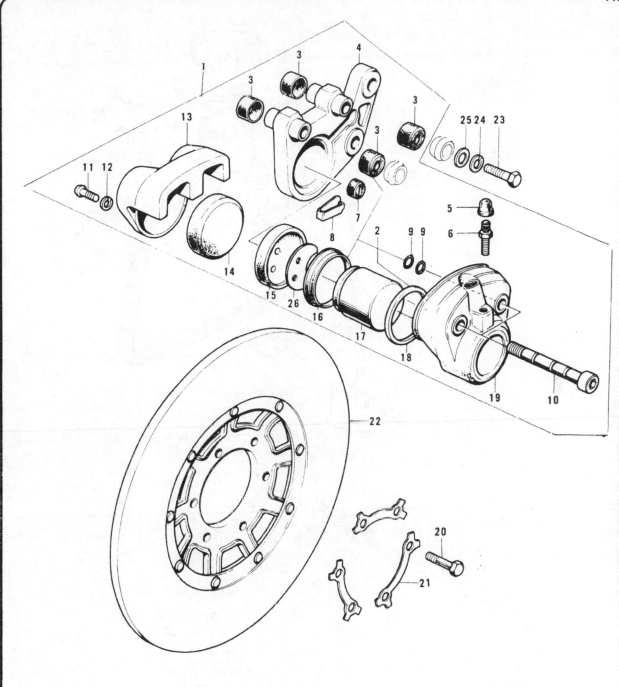

Fig. 5.3 Caliper assembly and disc

1	Caliper assembly	9	'O' ring - 4 off	17	Caliper piston
2	Caliper repair kit	10	Shaft - 2 off	18	Piston ring seal
3	Dust seal - 4 off	11	Pan head screw	19	Caliper 'A'
4	Caliper holder	12	Spring washer	20	Bolt - 6 off
5	Cap for bleeder nipple	13	Caliper 'B'	21	Lock washer - 3 off
6	Bleeder nipple	14	Disc pad 'B'	22	Disc plate
7	Rubber bush	15	Disc pad 'A'	23	Bolt - 2 off
8	Disc pad 'A' stopper	16	Piston seal	24	Spring washer - 2 off

25 Plain washer - 2 off
26 Shim

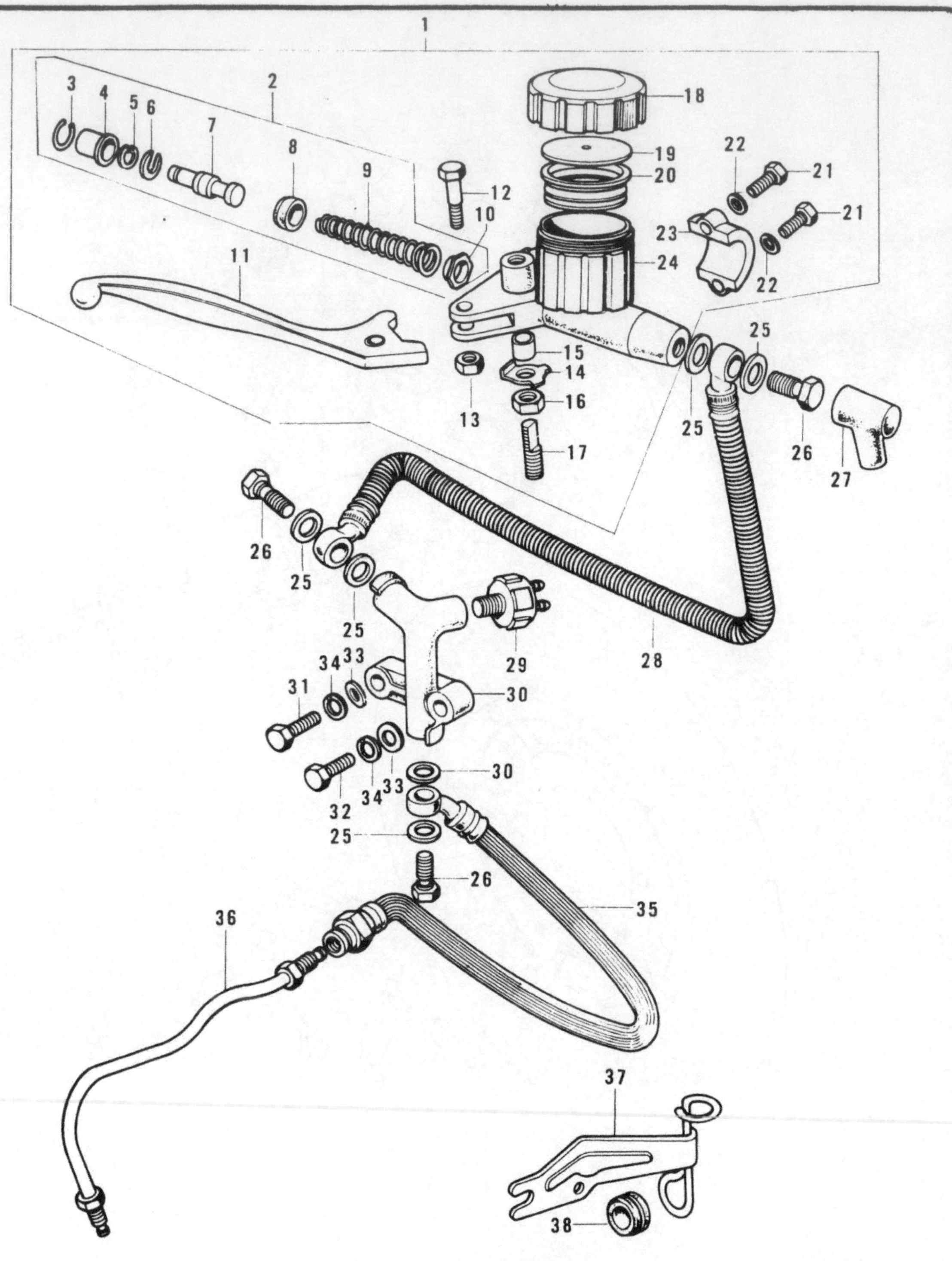

Fig. 5.4 Master cylinder and hoses

1 Master cylinder assembly	11 Front brake lever	21 Bolt - 2 off	31 Bolt
2 Repair kit	12 Lever bolt	22 Plain washer - 2 off	32 Bolt
3 Stopper	13 Nut	23 Master cylinder holder	33 Spring washer - 2 off
4 Boot	14 Lock washer	24 Body master cylinder	34 Plain washer - 2 off
5 Circlip	15 Tube	25 Oil bolt washer - 6 off	35 Hose
6 Piston stopper	16 Nut	26 Bolt - 3 off	36 Brake pipe
7 Piston	17 Lever adjusting bolt	27 Dust cover	37 Bracket
8 Primary cup	18 Oil reservoir cap	28 Hose	38 Hose grommet
9 Return spring assembly	19 Master cylinder plate	29 Front brake switch	
10 Check valve assembly	20 Diaphragm	30 Three way joint	

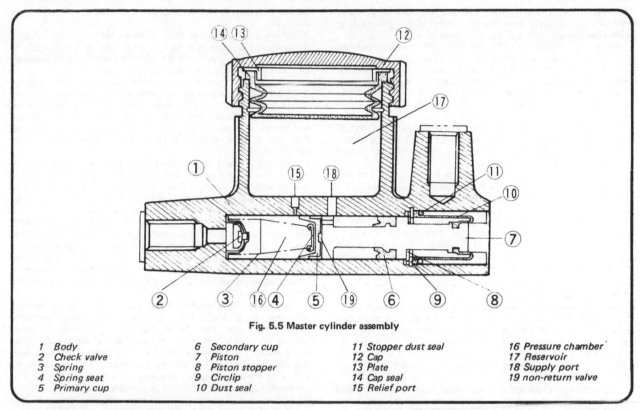

Fig. 5.5 Master cylinder assembly

1 Body	6 Secondary cup	11 Stopper dust seal	16 Pressure chamber
2 Check valve	7 Piston	12 Cap	17 Reservoir
3 Spring	8 Piston stopper	13 Plate	18 Supply port
4 Spring seat	9 Circlip	14 Cap seal	19 non-return valve
5 Primary cup	10 Dust seal	15 Relief port	

piston straight out without turning it. The piston can also be blown out with compressed air through the brake line outlet, if it is reluctant to leave its seat.

3 Remove the piston seal taking care not to damage the cylinder wall. All the remaining parts such as the bleeder valve and nipple, may be removed at this time.

4 Clean all the parts in cleaning fluid and then blow them dry, taking care to pass air through all the passages.

5 Inspect all the parts for a worn or damaged condition, and replace them as necessary. Check the brake friction pads for damage or excessive wear and renew them if necessary, especially if they have worn down to the red line. The pads should always be replaced as a pair. If the pads are oil impregnated, they should be renewed without question.

6 Inspect the oil seals for a worn, damaged or cracked conditon and renew them if necessary. If the seal around the caliper piston is bad, one pad will tend to wear more quickly than the other. The constant friction caused by the dragging pad, will cause a sharp increase in brake and fluid temperature, which may result in damage to the various parts.

7 When assembling the new seals and piston, coat them all liberally with clean brake fluid. The system is now ready for bleeding, as described in the following Section.

16 Bleeding the hydraulic brake system

1 A mushy feeling at the front brake lever can often be traced to air in the brake system. Brake fluid is not easily compressed, so that when the front brake lever is operated, almost all of the force applied to the lever is transmitted to the brake caliper. Air on the other hand compresses more easily, so that any air in the system compresses before the fluid will. This also means that some of the lever travel is wasted as it is used in compressing air without actually applying force to the caliper.

2 The brake system should be bled whenever the action at the front brake lever feels spongy, whenever brake fluid is changed, or whenever a brake line fitting has been loosened or disconnected. Bleed the hydraulic brake system in the following manner,

3 Remove the reservoir cap and check that there is plenty of fluid in the reservoir. The brake fluid level must be checked several times during the bleeding operation, and replenished as necessary. If the fluid in the reservoir runs out at any time, during the bleeding operation, air will enter the system and the procedure will have to be started all over again.

4 Use only new brake fluid. Do not use fluid that has been left unsealed in a container for any length of time. Use only the type of fluid recommended in this manual. Caution: Brake fluid must not be spilled on painted surfaces, it will strip the paint and should therefore be wiped off immediately.

5 Slowly pump the brake lever until no air bubbles can be seen rising up through the fluid from the holes at the bottom of the reservoir. When no more bubbles appear, the master cylinder end of the brake system has been purged of air.

6 Replace the reservoir cap, and run a clear plastic hose from the caliper bleeder valve into a container containing some brake fluid. The bottom of the hose must remain immersed throughout the whole operation. Unscrew the bleed nipple one turn and pump the front brake lever until no more air bubbles can be seen emerging from the plastic hose. Tighten the bleed nipple whilst the brake lever is squeezed towards the handlebars. During this process repeatedly check the fluid level in the reservoir, and replenish it as necessary.

7 When the system has been bled, replace the rubber cap on the bleeder valve, and check the fluid level in the reservoir once more. The handlebars must be turned so that the reservoir is level, and the brake fluid must come up to the level line inscribed inside the reservoir.

17 Wheel balancing and alignment

1 Wheel balancing should preferably be carried out on a truing stand, but as this necessitates removing the wheels, it can however be done in the machine. The front wheel can be balanced in

the forks as long as the brake is free and not binding. Similarly the rear wheel can also be balanced in the swinging arm as long as the chain is removed, and the rear brake is not binding.

Balancing should be accomplished with balance weights, although solder can be wrapped round the spokes as a substitute. The procedure is to spin the wheel in question and see where it stops, then mark the rim at its lowest point. Do this several times to determine where the heaviest section of the wheel is. Then begin adding weight to the opposite side of the rim as a counter-balance.

The weights available are measured in grams of 10, 20 and 30 or (1/3, 2/3 and 1 oz.). These can be obtained from a dealer. When the wheel is perfectly balanced it should not stop at any one point in particular, when spun.

2 Wheel alignment is easy on the Z1 series due to the fact that the rear fork ends are marked with a series of vertical lines. If the draw bolts are adjusted an even amount at a time as indicated by the lines, correct wheel alignment is preserved. If desired, the wheel alignment can be cross-checked by running a plank of wood parallel to the machine, so that it touches both walls of the rear tyre. If wheel alignment is correct, the measurement should be equal from either side of the front wheel tyre when tested on both sides of the rear wheel.

18 Tyres: removal and replacement

1 At some time or other the need will arise to remove and replace the tyres, either as a result of a puncture or because replacements are necessary to offset wear. To the inexperienced, tyre changing represents a formidable task, yet if a few simple rules are observed and the technique learned, the whole operation is suprisingly simple.

2 To remove the tyre from either wheel, first detach the wheel from the machine. Deflate the tyre by removing the valve core, and when the tyre is fully deflated, push the bead from the tyre away from the wheel rim on both sides so that the bead enters the centre well of the rim. Remove the locking ring and push the tyre valve into the tyre itself.

3 Insert a tyre lever close to the valve and lever the edge of the tyre over the outside of the rim. Very little force should be necessary; if resistance is encountered it is probably due to the fact that the tyre beads have not entered the well of the rim, all the way round.

4 Once the tyre has been edged over the wheel rim, it is easy to work round the wheel rim, so that the tyre is completely free from one side. At this stage the inner tube can be removed.

5 Now working from the other side of the wheel, ease the other edge of the tyre over the outside of the wheel rim that is furthest away. Continue to work around the rim until the tyre is completely free from the rim.

6 If a puncture has necessitated the removal of the tyre, re-inflate the inner tube and immerse it in a bowl of water to trace the source of the leak. Mark the position of the leak, and deflate the tube. Dry the tube, and clean the area around the puncture with a petrol soaked rag. When the surface has dried, apply rubber solution and allow this to dry before removing the backing from the patch, and applying the patch to the surface.

7 It is best to use a patch of the self vulcanizing type, which will form a very permanent repair. Note that it may be necessary to remove a protective covering from the top surface of the patch after it has sealed into position. Inner tubes made from a special synthetic rubber may require a special type of patch and adhesive, if a satisfactory bond is to be achieved.

8 Before replacing the tyre, check the inside to make sure that the article that caused the puncture is not still trapped inside the tyre. Check the outside of the tyre, particularly the tread area

to make sure nothing is trapped that may cause a further puncture.

9 If the inner tube has been patched on a number of past occasions, or if there is a tear or large hole, it is preferable to discard it and fit a replacement. Sudden deflation may cause an accident, particularly if it occurs with the front wheel.

10 To replace the tyre, inflate the inner tube for it just to assume a circular shape but only to that amount, and then push the tube into the tyre so that it is enclosed completely. Lay the tyre on the wheel at an angle, and insert the valve through the rim tape and the hole in the wheel rim. Attach the locking ring on the first few threads, sufficient to hold the valve captive in its correct location.

11 Starting at the point furthest from the valve, push the tyre bead over the edge of the wheel rim until it is located in the central well. Continue to work around the tyre in this fashion until the whole of one side of the tyre is on the rim. It may be necessary to use a tyre lever during the final stages.

12 Make sure there is no pull on the tyre valve and again commencing with the area furthest from the valve, ease the other bead of the tyre over the edge of the rim. Finish with the area close to the valve, pushing the valve up into the tyre until the locking ring touches the rim. This will ensure that the inner tube is not trapped when the last section of bead is edged over the rim with a tyre lever.

13 Check that the inner tube is not trapped at any point. Reflate the inner tube, and check that the tyre is seating correctly around the wheel rim. There should be a thin rib moulded around the wall of the tyre on both sides, which should be an equal distance from the wheel rim at all points. If the tyre is unevenly located on the rim, try bouncing the wheel when the tyre is at the recommended pressure. It is probable that one of the beads has not pulled clear of the centre well.

14 Always run the tyres at the recommended pressures and never under or over inflate. The correct pressures for solo use are given in the Specifications Section of this Chapter.

15 Tyre replacement is aided by dusting the side walls, parti-cularly in the vicinity of the beads, with a liberal coating of french chalk. Washing up liquid can also be used to good effect, but this has the disadvantage of causing the inner surface of the wheel rim to rust.

16 Never replace the inner tube and tyre without the rim tape in position. If this precaution is overlooked there is a good chance of the ends of the spoke nipples chaffing the inner tube and causing a crop of punctures.

17 Never fit a tyre that has a damaged tread or sidewalls. Apart from legal aspects, there is a very great risk of a blowout, which can have very serious consequences on a two wheeled vehicle.

18 Tyre valves rarely give trouble, but it is always advisable to check whether the valve itself is leaking before removing the tyre. Do not forget to fit the dust cap, which forms an effective extra seal.

19 Tyres: valves and dustcaps

1 Inspect the valves in the inner tubes from time to time making sure the seal and spring are making an effective seal. There are tyre valve tools available for clearing damaged threads in the valve body, and incorporating thread cleaning for the outside thread of the body. A key is also incorporated for tightening the valve core.

2 The valve caps prevent dirt and foreign matter from entering the valve, and also form an effective second seal so that in the event of the tyre valve sticking, air will not be lost.

Tyre changing sequence - tubed tyres

 Deflate tyre. After pushing tyre beads away from rim flanges push tyre bead into well of rim at point opposite valve. Insert tyre lever adjacent to valve and work bead over edge of rim.

Use two levers to work bead over edge of rim. Note use of rim protectors

 Remove inner tube from tyre

When first bead is clear, remove tyre as shown

 When fitting, partially inflate inner tube and insert in tyre

Work first bead over rim and feed valve through hole in rim. Partially screw on retaining nut to hold valve in place.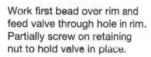

 Check that inner tube is positioned correctly and work second bead over rim using tyre levers. Start at a point opposite valve.

Work final area of bead over rim whilst pushing valve inwards to ensure that inner tube is not trapped

20 Fault diagnosis: wheels, brakes and tyres

Symptom	Cause	Remedy
Handlebars oscillate at low speed	Buckle or flat in wheel rim, most probably front wheel	Check rim alignment by spinning the wheel. Correct by retensioning spokes or having wheel rebuilt on new rim. Check tyre alignment
	Tyre not straight on rim.	
Machine lacks power and accelerates poorly	Brakes binding Wrongly adjusted caliper.	Hot brakes drums provide best evidence. Readjust caliper.
Brakes grab when applied gently	Ends of brake shoes not chamfered (on internal expanding brakes)	Chamfer with a file.
	Eliptical brake drum Faulty caliper, on disc brake Warped disc.	Lightly skim in a lathe by a specialist. Replace with a new caliper. Replace disc if beyond skimming limit.
Brake squeal	Glazed pads	Lightly sand the pads, and use the brake gently for a hundred miles or so until they have a chance to bed in properly.
	Improperly adjusted caliper	Readjust caliper, as described in section 15 of this chapter.
	Extremely dirty and dusty front brake caliper and disc assembly	Clean with water, do not use high pressure spray equipment.
Excessive lever travel on front brake	Air in system, or leak in master cylinder or caliper; worn disc pads.	Bleed the brake Replace the cylinder seals Replace the pads.

Chapter 6 Electrical system

Contents

General description 1	Headlamp: replacing bulbs and adjusting beam height ... 10
Crankshaft alternator: checking the output 2	Stop and tail lamps: replacing bulbs 11
Battery: maintenance procedure 3	Flashing indicator relay and lamps: Location and replacement 12
Battery: charging procedure 4	Speedometer and tachometer: replacing bulbs 13
Silicon rectifier: location, testing, and replacement 5	Horn location and adjustment 14
Voltage regulator: operating principle and testing 6	Handlebar switches, ignition and lighting switches:
Fuse: location and replacement 7	examination replacement 15
Starter motor: removal, examination, and replacement ... 8	Stop lamp switches: adjustment and replacement 16
Starter motor switch: function and location 9	Engine oil pressure switch: removing and replacement ... 17
	Fault diagnosis: electrical system 18

Specifications

Battery
Make	Yuasa
Type	12 N 14-3A
Voltage	12 volt
Amp hour capacity	14 amp hour
Earth or ground	Negative

Generator/Alternator:
Make	Kokusan
Type	AR 3701
Output	12 volts

Ignition coil
Make	Kokusan
Type	1G 3303, 1G 3304

Starter motor:
Make	Kokusan
Type	SM 226 K
Brush length new	½ inch (12 — 13 mm)
Minimum permissable	¼ inch (7 mm)

Bulbs:
Headlamp	12V 50/35W
Stop/tail lamp	12V 4/32W or 8/23W
Neutral indicator lamp	12V 3.4W
Instrument lamps	12V 3.4 W
Indicator lamps	12V 3.4W
Flashing direction lamps	12V 23W x 4

Fuse ... ... 20 amp

1 General description

1 The Z1 models covered by this manual use a 12 volt electrical system. The system comprises a crankshaft driven AC generator of the rotating magnet type surrounded by a stator coil assembly.

2 The output of the generator is AC hence the need for a rectifier of the silicon type to convert the current to DC so that it can be used to charge the battery. The standard output with a fully charged battery is 15 to 20 volts.

2 Crankshaft alternator: checking the output

1 The output from the alternator can be checked by connecting both a voltmeter and an ammeter into the battery circuit. Refer to Section 2 of Chapter 3 for details. The test gives only an approximate indication as to whether the alternator is functioning correctly within its prescribed limits. It is necessary to seek the assistance of a Kawasaki Agent or an Auto Electrical Expert to determine whether the generator is working to peak efficiency. The average rider/owner is unlikely to have the necessary test equipment available to carry out a sufficiently accurate check.

5.2 Rectifier mounted on panel

3 Battery: maintenance procedure

1 The battery fitted to the Z1 models is a Yuasa 12 volt with a 14 amp. hour capacity.

2 The transparent case of the battery gives an instant visual check of the level of electrolyte contained in the battery. The battery fits under the dualseat in a special cradled compartment.

3 The level of the electrolyte in the battery should never be allowed to fall below the lower level mark. If it does, it must be topped up with distilled water to the upper level, after the initial fill with sulphuric acid of a specific gravity of 1,260 to 1,280. Also make sure the vent pipe is routed through the proper channel provided to ensure that it discharges clear of the frame parts.

4 It is seldom practicable to repair a cracked battery case because the acid that is already established in the crack will prevent the formation of an effective seal. A cracked battery should be renewed at once because apart from a deterioration in efficiency there will be a considerable amount of corrosion if the acid continues to leak.

5 The battery should be checked every week, and topped up when necessary to the upper electrolyte level. When the machine is laid up for any length of time, it is always advisable to remove the battery and give it a refresher charge every four to six weeks, using a battery charger. Once a battery has been put into service (filled with acid) it must be kept in use, otherwise the cell plates will sulphate and render it useless.

8.1 Undo the mounting bolts to free starter motor

4 Battery: charging procedure

1 The normal charging rate for batteries of up to 14 amp. hour capacity is 1½ to 2 amps. It is permissible to charge at a more rapid rate in an emergency but this shortens the life of the battery, and should be avoided. Always remove the vent caps when recharging a battery, otherwise the gas created within the battery when charging takes place will explode and burst the case with disastrous consequences.

5 Silicon rectifier: location and replacement

1 The silicon rectifier fitted to the electrical system converts the AC current produced by the alternator to DC so that it can be used to charge the battery.

2 The rectifier is mounted on the electrical panel with the voltage regulator and connector blocks, on the right-hand side of the machine below the dualseat. The whole panel is enclosed by

10.1 3 Screws retain rim

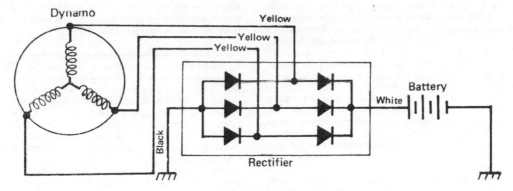

Fig. 6.1 Rectifier circuit

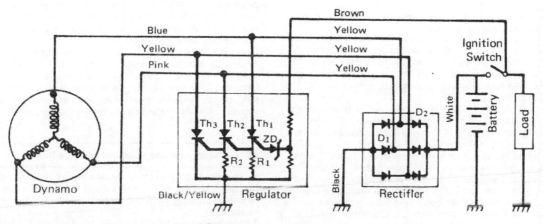

Fig. 6.2 Rectifier and regulator circuit

a cover that is pushed on to fasteners.

3 The rectifier is a component that cannot be repaired, and if found faulty it has to be replaced with a new unit. Damage to the unit can be caused by running the machine without a battery or if the battery leads are accidentally reversed.

4 The six-diode arrangement (two diodes for each of the dynamo's three output phases) is used to convert the AC current into DC current for battery charging, ignition, lighting, and horn circuits. The diodes in the rectifier can only conduct current from negative to positive, and therefore they convert AC to DC. If the rectifier or diodes become faulty they will conduct current either in both directions, or not at all, and therefore lead eventually to a discharged battery.

5 The rectifier can be tested with an ohmmeter. First disconnect the white rectifier plug from the connector panel, and the white lead going to the battery. With the tester set on the RX10, or RX100 range, check the resistance between the white rectifier lead and each of the yellow leads, the yellow leads and the white lead, the black lead and each of the yellow leads, and also each yellow and black lead. This involves a total of twelve measurements. The resistance should be low in one direction, and about ten times as great in the other direction. If the readings are the same in either direction for any pair of wires, the rectifier is faulty and should be replaced. The lower reading should be within the 1/3rd scale of zero ohms, regardless of the type of tester used.

6 Note: When removing or installing a rectifier, do not loosen or try to tighten the main assembly nut of the rectifier, as this is part of the assembly and should not be disturbed. If disturbed damage will be caused to the whole rectifier assembly and render it useless. When fitting a new replacement take great care

not to disturb the coating over the electrodes, which may peel or flake and destroy the working action.

6 Voltage regulator: operating principle and testing

1 The voltage regulator fitted to the Z1 models is of the solid state type. Its function is to handle the power output from the three phase generator, and to limit the voltage to 15 - 16 volts. It is constructed to control each of the three phases of the alternator output.

2 Two symptoms which would indicate the possibility of a faulty regulator are repeated battery discharging or battery overcharging. A battery overcharged is indicated by the need to top up the electrolyte more frequently than is normal, and also by blowing bulbs in the lighting system when running at high rpm's.

3 Discharging of the battery more excessively than is normal is indicated by a battery that when checked reads correctly, but goes dead quickly after being fully charged.

4 To test the voltage regulator disconnect the green regulator plug from the connection panel. This is located under the right-hand plastic cover. When removing the regulator do not loosen or remove the three screws contained in the regulator body. These screws aid heat dissipation; the unit will overheat badly if they are not properly installed.

5 With an ohmmeter set on the RX10 or RX100 scale, there should be 1,000 - 1,100 ohms resistance between the black and brown leads, and no reading (00) between any other lead. Any other results indicate that the regulator is defective.

6 For the next test a 14 V.D.C. and a 16-17 V.D.C. power

source must be available. If the sources used cannot provide sufficient power, the tests will be inaccurate, and if more than 18 volts are passed through the regulator it may be damaged.

7 Connect the regulator to the 14 V.D.C. source, set the ohmmeter to either the RX10 or the RX100 range, and check that there is no reading (00) between the black lead and pink lead or the yellow lead and blue lead. If the meter gives any reading for any of these leads, the regulator is defective.

8 Connect the regulator to the 16-17 V.D.C. source in the same manner as paragraph 7, and set the meter on the RX1 scale, this should result in a very low reading when the meter is connected between the black lead and the blue, pink or the yellow leads. If there is no reading between any or all of the leads, or if any one reading is higher than the other two, the regulator is defective and must be replaced with a new one.

Note: If the voltage source and regulator are connected backwards for even a moment, the regulator will be damaged. Also be sure that the black and brown leads never touch the meter leads at the same time, or the meter will be damaged.

7 Fuse: location and replacement

1 A fuse is incorporated in the electrical system. It is contained in a plastic holder and clips to the side of the battery holder. The fuse is incorporated in the system to give protection from a sudden overload such as could happen with a short circuit. The fuse is rated at 20 amps.

2 If the fuse blows it should not be renewed until the cause of the short is found. This will involve checking the electrical circuit to correct the fault. If this rule is not observed, the fuse will almost certainly blow again.

3 When a fuse blows and no spare is available a "get you home" remedy is to wrap the fuse in silver paper before replacing it in the fuse holder. The silver paper will restore electrical continuity by bridging the broken wire within the fuse. Replace the doctored fuse at the earliest opportunity to restore full circuit protection. Make sure any short circuit is elimimated first.

4 Always carry two spare fuses of the correct rating.

8 Starter motor: removal, examination and replacement

1 An electric starter is fitted to the Z1 models operated by a push button switch on the body of the right-hand twistgrip. The starter motor is mounted on the top of the crankcase, to the rear of the cylinder block, and is accessible by removing the cover that is held on by two screws. The motor is held by two bolts that go through the end bracket and secure the motor to the crankcase. Remove the fuel tank, carburettors, starter cover and gasket, also the left-hand engine cover. Remove the right-hand side cover, unscrew the starter wire from the starter relay terminal, and free the wire from the cable clamp. Unscrew the two mounting bolts and the starter motor can now be removed.

2 To dismantle the motor, first remove the two long assembly screws through the body. The end cover can then be removed. Disconnect the brush assembly from the field coil lead, and then remove the brush plate and brushes.

3 Remove the remaining end cover, and withdraw the armature from the starter body and field coil assembly. Inspect the carbon brushes for wear, and replace them if they have worn past the standard measurement of ½ inch (12 to 13 mm) to ¼ (7 mm) or more. Always replace the brushes as a pair.

4 Before fitting the brushes make sure the commutator is clean. Clean with a strip of fine emery pressed against the commutator whilst it is rotated by hand, then wipe with a clean rag soaked in petrol.

5 Reassemble the starter motor by reversing the dismantling procedure, place it back into the top of the crankcase and tighten the two mounting bolts. Refit the starter cable to the switch and replace the plastic cover. Replace the starter motor cover with the gasket and two screws.

Note: When reassembling the starter motor make sure the marks

scribed on the body are in line, also when slipping the starter motor into place lubricate the "O" ring with a small amount of fresh oil. This facilitates reassembly.

9 Starter motor switch: function and location

1 The starter motor switch is designed to work on the electromagnetic principle. When the starter motor button is depressed, current from the battery passes through the windings of the solenoid switch and generates an electro-magnetic force which causes a set of contacts to close. Immediately the points close, the starter motor is energised and a very heavy current is drawn from the battery.

2 This arrangement is used for two reasons. First the current is drawn only when the starter button is depressed and is cut off again when pressure on the button is released. This ensures minimum drainage to be taken from the battery. Secondly, if the battery is in a low state of charge, there will not be sufficient current to cause the solenoid contacts to close. In consequence, it is not possible to place an excessive drain on to the battery, which in some circumstances can cause the battery plates to overheat and shed their coatings.

3 If the starter will not operate, first suspect a discharged battery or a defective switch. If there is a "click" when the button is depressed the solenoid switch is functioning but the battery is probably discharged. If there is no contact, the solenoid switch probably needs renewing (provided the circuit is correct). It is not practicable to repair a solenoid switch.

10 Headlamp: replacing bulbs and adjusting beam light

1 In order to gain access to the headlamp bulbs remove the rim, this is retained by two screws behind the rim. The rim can now be pulled off with the reflector unit complete and the pilot bulb removed.

2 Disconnect the headlamp bulb adaptor from the sealed beam unit and remove the lens retaining ring. The main bulb can now be removed.

3 To adjust headlight beam height, slacken the two turn signal mounting nuts inside the headlamp shell, loosen the mounting bolts underneath the lamp and adjust the aim of the unit to the required position (up and down) vertical aim.

4 Adjust the horizontal (left to right) aim of the light by turning the small crosshead screw situated directly in front of the rim. Screwing the screw inwards moves the beam to the right and screwing out moves the beam to the left. On European models the headlamp bulb is the prefocus type, on USA models the headlamp lens and bulb are a sealed unit, and the whole unit has to be replaced in the event of light failure. Set the beam height with the machine on a level surface 25 yards from a wall so that the centre of the light spot is the same distance as that from the centre of the headlamp to the ground.

11 Stop and tail lamp: replacing the bulb

1 The tail lamp fitted to the Z1 series has a double filament bulb. One lights the rear lamp and the other for the braking of the machine. The brake light is operated by either the front or rear brake lamp switch. The front brake switch is an oil pressure switch that is installed in the front hydraulic brake hose and it turns on when front brake pressure is applied. The rear brake switch is operated by the rear brake pedal, and is adjustable by altering its position higher or lower in the mounting bracket.

2 Remove the two long screws that retain the rear lamp lens. The bulb can be removed by pushing in and at the same time turning in an anticlockwise direction. Replace the bulb by reversing the procedure. The bulb has to be renewed if either the tail lamp or brake light filament burns out. When the lens is

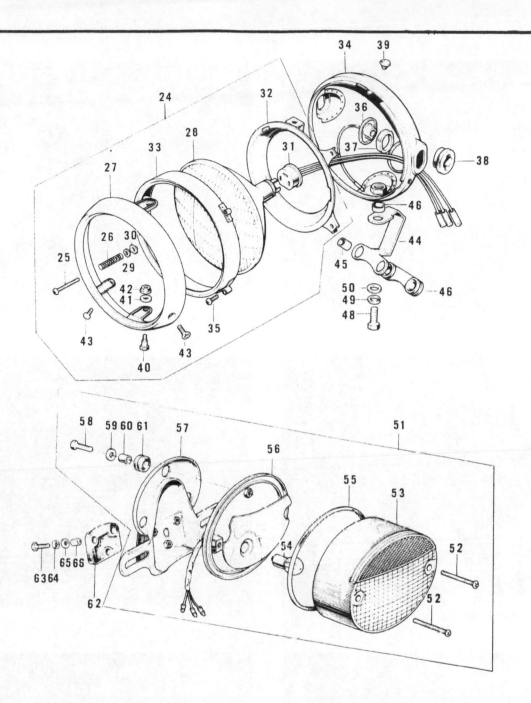

Fig. 6.3 Headlamp and tailamp assembly

24 Headlamp assembly
25 Adjusting screw
26 Adjusting screw spring
27 Headlamp rim
28 Sealed beam unit
29 Plain washer
30 Nut
31 Socket
32 Ring
33 Ring
34 Headlamp shell

35 Pan head screw - 2 off
36 Collar - 2 off
37 Rubber for headlamp shell
38 Rubber for headlamp shell
39 Rubber plug
40 Pan head screw - 2 off
41 Plain washer - 2 off
42 Nut - 2 off
43 Pan head screw - 2 off
44 Bracket
45 Collar - 2 off

46 Rubber
47 Collar
48 Bolt
49 Spring washer
50 Plain washer
51 Tail lamp assembly
52 Pan head screw - 2 off
53 Tail lamp lens
54 Bulb stop and tail
   12 v 32/4 watt
55 Rubber gasket

56 Body tail lamp
57 Tail lamp and
   plate bracket
58 Bolt - 3 off
59 Plain washer - 3 off
60 Collar - 3 off
61 Rubber - 3 off
62 Shock damper
63 Bolt - 3 off
64 Plain washer - 3 off
65 Collar - 3 off
66 Rubber - 3 off

10.1A Reflector unit removed complete

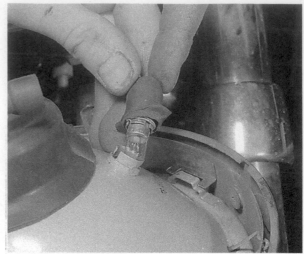

10.1B Take out pilot bulb

10.2 Disconnect adaptor

10.2A Remove cap

10.2B Take out bulb

11.2 Removing rear lamp lens

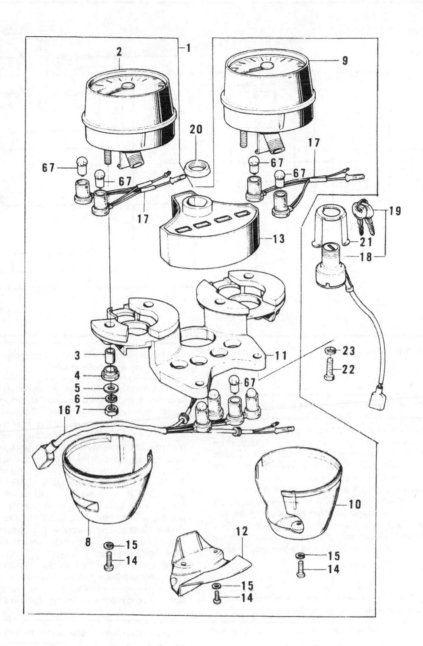

Fig. 6.4 Speedometer and tachometer assembly

1 Speedometer and tachometer assembly combined	7 Nut	14 Pan head screw - 5 off	19 Key set
2 Speedometer only	8 Cover	15 Spring washer - 5 off	20 Locknut
3 Meter collar - 4 off	9 Tachometer assembly	16 Socket assembly - indicators	21 Igniton switch holder
4 Rubber - 4 off	10 Tachometer cover	17 Socket assembly meter lamp - 2 off	22 Bolt - 2 off
5 Plain washer - 4 off	11 Bracket	18 Ignition switch	23 Washer - 2 off
6 Spring washer - 4 off	12 Wiring harness cover		67 Bulb 12v. 3 watt
	13 Indicator lamp cover		

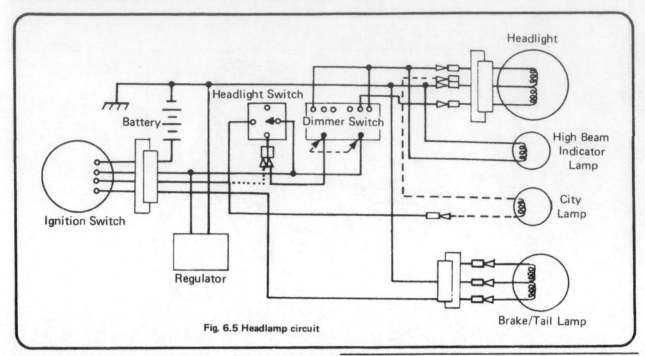

Fig. 6.5 Headlamp circuit

replaced, make sure the mounting gasket is in good condition and waterproof.

12 Flashing indicator relay and lamps: location and replacement

1   The flashing indicator relay is fitted to the same electrical panel as the voltage regulator and rectifier, below the dualseat on the right-hand side of the machine. It is mounted in rubber because of the fragile mechanism inside. It is very important not to drop the unit otherwise damage will result.

2   The flashing indicator lamps are fitted to the front and rear of the machine on 'stalks' through which the wires pass. To renew the bulbs remove the two screws that retain each lens, and remove the bayonet type bulbs. These are single filament with a rating of 12 volt 23 watt. Make sure the rubber gaskets on the base of the lens are in good condition and waterproof, when replacing the lens.

13 Speedometer and tachometer: replacing the bulbs

1   The bulbs that fit into the instruments and dash panel are of the small bayonet type, rated at 12 volts 3.4 watt.

2   The shrouds that cover the instruments have to be removed to expose the bulbs; the bulb holders are a push fit into the back of the panel and are easily removed and replaced.

14 Horn: location and adjustment

1   The horn is adjustable by means of the small screw located at the back of the horn, situated in the top of the front frame gusset. To adjust the volume, turn the screw about half a turn either way until the desired tone is required.

2   If it is necessary to dismantle the horn to clean the contacts, first remove the fuel tank, then remove the horn. Clean the contacts with a fine sand paper, and if after this the horn does not work it must be renewed. Make sure the horn is watertight by renewing the gasket when reassembling.

15 Handlebar switches, ignition and lighting switches: examination and replacement

1   The handlebar switches are made up of two halves that clamp together with small crosshead screws, these are situated underneath the switch assemblies. The switches seldom give any trouble, and it is not advisable to take them apart as the parts are so small that difficulties can occur during reassembly, not to mention the time involved. If a switch fails it is far better to fit a new replacement. This is quite a simple task as the wires are fitted into snap connectors.

2   The main ignition switch is located in the centre of the dash panel and is removed by unscrewing the ring nut round the barrel of the switch. The dash panel can then be removed. Take off the front light unit and headlamp shell together with the flasher lamps and remove the switch lower cover and the mounting nut. The ignition switch can now be removed by unplugging the leads.

3   When replacing the ignition switch, the vertical aim of the headlight will have to be readjusted. Also note that the left-hand turn signal wires goes to the green wire, and the right-hand turn signal wire is plugged into the grey wire.

16 Stop lamp switches: adjustment and replacement

1   The rear brake stop lamp switch is located in a bracket above the rear brake pedal and is operated by an expansion spring linked to the rear brake pedal. The body of this switch is threaded to enable it to be raised or lowered.

2   If the rear brake stoplamp is late in operating, slacken the two locknuts and raise the switch body. When the adjustment is correct tighten the locknuts and test. If the stoplamp is early in operation, slacken the locknuts and lower the body in relation to the bracket.

3   As a guide the light should come on when the rear brake pedal has been depressed about ¾ inch (2 cm).

4   The hydraulic front brake lamp switch operates the same bulb in the tail light as the rear brake switch. The hydraulic pressure switch switch operates when the front brake lever is compressed. Adjustment of this switch is not possible. If the pressure switch has to be renewed, the complete switch can be unscrewed after the hydraulic system has been drained from the

12.2 Removing indicator lens

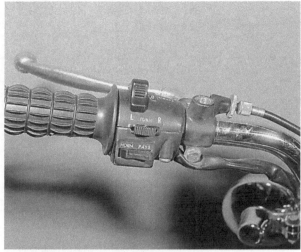
15.1 Switches are clamped with two screws

16.1 Rear stop lamp switch ....

16.1A ....switch is adjusted by two nuts

17.1 The oil pressure switch

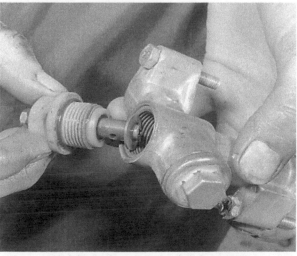
17.1A Removing pressure switch from body

three way branch pipe situated underneath the headlamp. When a new hydraulic pressure switch is installed it will be necessary to refill and bleed the hydraulic brake system as described in Section 16, Chapter 5.

17 Engine oil pressure switch: removing and replacement

1   The oil pressure switch is mounted on top of the engine to the rear of the cylinder block. The oil pressure switch serves to indicate when the oil pressure has dropped due to pump failure, blockage in a oilway or too little oil available to the oil pump. It

is not however intended to be used as an indication of correct oil level. If the oil pressure lamp (located in the dash panel) comes on and stays on when the oil is hot and the machine is being rapidly accelerated, the fault is probably the switch, this can sometimes be remedied by revving the engine up past 6.000 rpm for a second or two but if this does not put the light out, disconnect the blue wire from the oil pressure switch, and remove the switch. When installing a new switch coat the thread with a sealer to form an oil tight seal. MAKE SURE IT IS THE SWITCH AT FAULT BEFORE USING THE MACHINE. A GENUINE LUBRICATION PROBLEM WILL CAUSE SEVERE ENGINE DAMAGE.

18 Fault diagnosis: electrical system

Symptom	Cause	Remedy
Complete electrical failure	Blown fuse	Check wiring for loose connections before fitting a new fuse.
	Isolated battery	Check battery connections for signs of corrosion.
Constant blowing of bulbs	Vibration or poor earth connections	Check bulb holders, check earth return connections.
Dim lights, horn and starter do not work	Discharged battery	Recharge battery with a battery charger. Check generator for output.
Starter motor sluggish or will not work	Worn brushes	Remove starter motor and replace with new bush. Clean commutator.
Flashing lights will not flash	Faulty relay unit	Replace with a new relay unit.
	Bad earth	Check flasher lamp bulb holders for good earth.

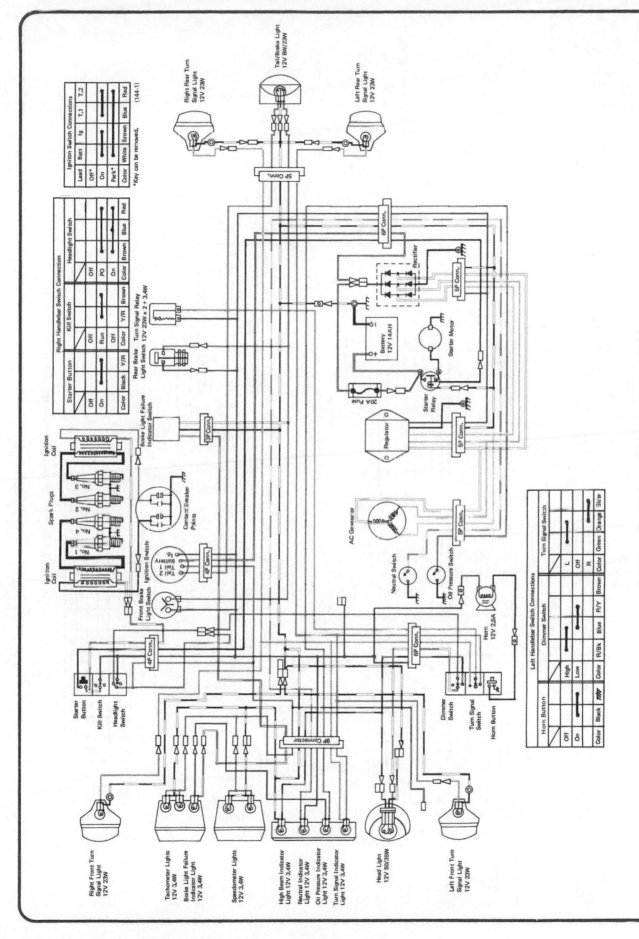

Fig. 6.6 Kawasaki 900cc Z1 Wiring diagram (U.S. model)

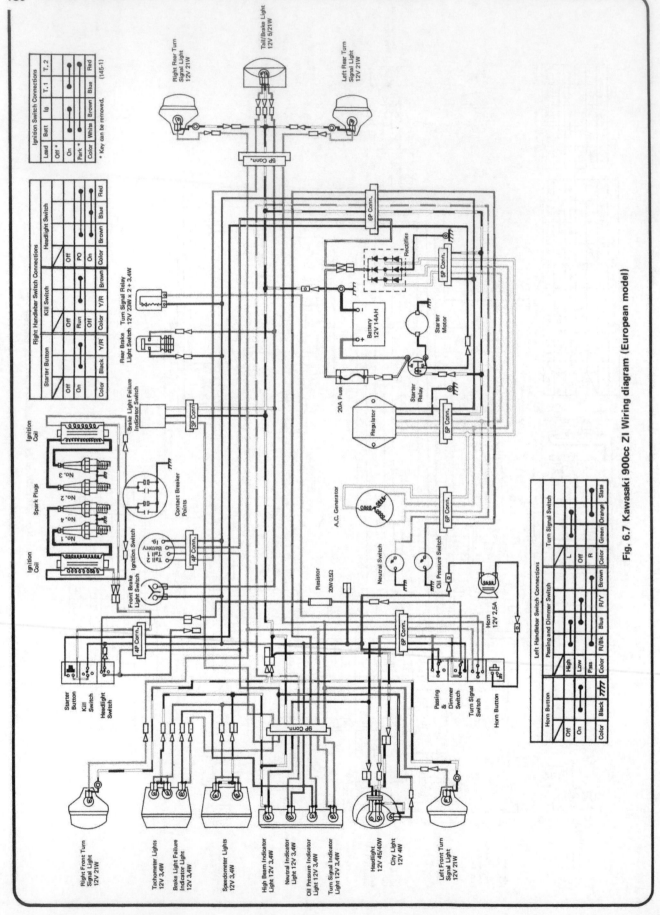

Fig. 6.7 Kawasaki 900cc Z1 Wiring diagram (European model)

Left and right-hand views of the 1977 Kawasaki Z1000 DOHC

Chapter 7 The Z900 and Z1000 models

Contents

General description
Swinging arm fork: removal and renovation 1
Final drive chain: maintenance and renewal 2
Hydraulic disc brakes: modifications 3
Rear brake master cylinder: maintenance, removal and renovation 4
Rear brake caliper: maintenance, removal and replacement ... 5

Specifications

The additional specifications given here cover the Z900 and Z1000 models where they differ from those given in the main text for the Z1 machines. Where specific mention is not made, it may be assumed that the Z1 specifications can be applied.

	Z900	Z1000
Engine		
Bore	66 mm	70 mm
Stroke	66 mm	66 mm
Displacement	903 cc	1015 cc
Compression ratio...	8.5 : 1	8.7 : 1
Maximum horsepower	81 @ 8,000 rpm	83 @ 8,000 rpm
Maximum torque...	7.3 kg/m @ 7,500 rpm	8.1 kg/m @ 6,500 rpm
Piston ring end gap	0.2 - 0.4 mm	0.3 - 0.5 mm
	(0.008 - 0.016 in)	(0.012 - 0.020 in)
Carburettors		
Make	Mikuni	Mikuni
Type	VM 26 SS	VM 26 SS
Main jet	115 - R	107 - 5R
Pilot jet	17.5	17.5
Needle jet	0 - 6	0 - 8
Jet needle...	5DL31 - 3	5CN8 - 3
Needle position	No. 3 groove	No. 3 groove
Tyre pressures		
Front...	28 psi	28 psi
Rear solo	32 psi	32 psi
Rear pillion	36 psi	36 psi
Front forks		
Spring free length service limit	455 mm (17.91 in)	455 mm (17.91 in)
Oil capacity at oil change (per leg)	approx. 140 cc	approx. 140 cc
Oil capacity after overhaul (per leg)	170 - 178 cc	170 - 178 cc
Oil level (from top of stanchion, with fork extended)	426 mm (16.77 in)	426 mm (16.77 in)
Oil type	SAE 10	SAE 10
Brakes		
Front (UK)	Twin disc	Twin disc
Front (US)	Single disc	Single disc
Rear	Disc or drum	Single disc
Battery		
Make	Yuasa	Yuasa
Type	YB 10	YB 14L-AZ
Capacity	12V 10AH	12V 14AH
Generator		
Make	Kokusan	Kokusan
Type	AR 3703	AR 3703
Starter motor		
Make	Mitsuba	Mitsuba
Type	SM-226-K	SM-226-K
Ignition coil		
Make	Toyo Denso	Toyo Denso
Type	ZC001-14/ZC001-23	ZC001-14/ZC001-23

General description

In 1976, the highly successful Z1 models were superseded by the Z900. The updated specification included audible turn signal indicators, now deleted, hazard warning flashers and twin front disc brakes, the latter having been an option for some time.

The Z900, although equally successful, produced marginally less power than its earlier ancestors, mainly due to the addition of the PCV system. Although the effect of this was, to all practical purposes, academic, Kawasaki were quick to introduce the Z1000 model later in the same year.

The Z1000 produced even more power than the original Z1, this being obtained by increasing the bore size from 66mm to 70mm. The exhaust system had by now been modified to a sleeker four-into-two arrangement. Many subtle alterations were incorporated into the new model.

Many of the engine castings were thickened to give quieter running. A heavier-webbed crankshaft made the engine smoother, despite it being over-square in configuration, so that low speed tractability was maintained.

The frame also received attention, being gusseted and reinforced at various points. The rake of the steering head was reduced slightly, and a new lengthened swinging arm fitted, running on needle roller bearings.

Most of the model changes and modifications have no real effect on the methods described in the foregoing text concerning maintenance and renovation: The manufacturers have preserved the initial concept of the Z1 wherever possible, and have restricted all subsequent alterations to the minimum, their policy being one of refinement, rather than change for change's sake.

This Chapter then, deals with those operations which demand a different approach or method when dealing with a specific task or component. Where specific mention is not made, it may be assumed that any modifications do not materially affect the methods of operation described in the preceding Chapters.

1 Swinging arm fork: removal and renovation

1 As mentioned in Section 1 of this Chapter, the swinging arm on Z1000 models is supported by needle roller bearings, rather than by the bushes used on earlier models. The use of needle rollers gives a greater degree of precision than the earlier bushed type, and they are also less prone to wear.

2 The bearings can be checked for wear by pushing the swinging arm fork from side to side. This is best done whilst the rear wheel is removed. Another good indication that all is not well is a slight twitch, particularly evident when the machine is ridden hard through a series of bends.

3 Access to the bearings is gained after removing the swinging arm as described in Section 2, paragraphs 9 to 17. It should be noted at this juncture that it is not possible to remove the bearings without damaging them, so be sure to have new bearings to hand before commencing work.

4 It is advisable to remove the brake torque arm from the swinging arm as it tends to get in the way. It is retained by a spring pin, nut and spring washer.

5 Withdraw the inner sleeve spindle, which forms the inner bearing race, and through which the pivot shaft passes. Examine the sleeve spindle for wear at the bearing surfaces, looking for pitting or grooves which may have occurred due to lack of lubrication or old age. If the surface is anything less than perfect, the sleeve should be renewed. It is false economy to re-use a worn sleeve, as the damaged surfaces will quickly destroy the new bearings.

6 Using a long drift, such as a long bolt, pass it through the bore of the swinging arm and drive out the bearing on the opposite side. Turn the swinging arm over and drive out the other bearing.

7 The new bearings are best fitted by using the drawbolt method. Obtain three plain washers and a spacer which will fit

over the pivot shaft. Check that when assembled the nut can be tightened to hold the assembly firmly against each end of the swinging arm bore. Fit a washer behind the pivot shaft head, and pass the shaft through the swinging arm bore. Slide on a bearing, followed by another washer, the spacer, the third washer, and finally the nut. Ensure that the washer adjacent to the bearing is larger in diameter than the bearing, and that the assembly, particularly the bearing, is square in relation to the bore.

8 Gradually tighten the nut so that the bearing is drawn squarely into the swinging arm bore. Remove the drawbolt components and repeat on the other side. Always lubricate the new bearings with gear oil before fitting.

9 Replace the swinging arm in the frame, following the removal sequence in reverse. Do not omit to loop the endless chain over it before refitting the wheel. Before using the machine, lubricate the pivot and bearing with high melting point grease. Pump the grease in until it just begins to ooze from the ends of the swinging arm bore.

10 Re-adjust the rear chain and brake before taking the machine on the road. The torque arm nut should be tightened to 2.6 - 3.5 kg m (19.0 - 25.0 ft lbs) and the spring pin replaced.

2 Final drive chain: maintenance and renewal

1 With the exception of the Z1 and Z1A models, which were fitted with a chain lubrication pump, the later models are supplied with what the manufacturers describe as pre-lubricated chains. These consist of roller chains with small O-rings at the ends of each roller which retain the lubricant that is applied on assembly.

2 Without exception, the chains are of the endless type. Spring links are not fitted as they are not considered strong enough to take the tremendous power produced by the engine. Consequently, it is not advisable to attempt to separate and re-rivet the chain, should it need removal, as this will probably weaken it considerably.

3 Despite its title, it should not be assumed that the chain does not need periodic maintenance. On the contrary, in view of the fact that considerable dismantling is necessary to renew the chain, frequent cleaning and lubrication will prolong its life and postpone the inevitable.

4 The chain should be cleaned in situ, and relubricated with one of the proprietary aerosol lubricants designed for this purpose. Engine oil is not really suitable as it tends to be flung off easily. Do not use paraffin as a cleaner and do not over-lubricate. Little and often is advised.

5 This task should be undertaken every 200 miles at least. In wet, dirty conditions this period can be reduced considerably, lubrication being especially important before and after a long run.

6 It is unfair to expect the rear chain to last for more than a year. In practice it is not usual to have to renew the chain more frequently, especially if the machine is driven hard.

7 Chain wear can be measured with the chain stretched taut by means of the chain tension adjusters. Measure the length of 20 links on the top run of the chain. It is easier to remove the chainguard first. The distance of a new chain, pin centre to pin centre, should be 381mm (15 in). If the distance has increased to more than 388mm (15.27 in), the chain should be renewed.

8 The correct way to remove the chain is as follows:

9 Place the machine securely on its centre stand, and block, if necessary, to raise the rear wheel clear of the ground. The silencers should be detached to give better access, as should the chain guard.

Drum brake models

10 Slacken off both chain adjusters, and the wheel spindle nut. Allow the adjusters to drop, exposing the stops at the end of the fork ends. These are each retained by a single bolt and should be removed.

11 Push the wheel forward so that the chain can be disengaged from the sprocket. Allow it to rest to one side. Disconnect the

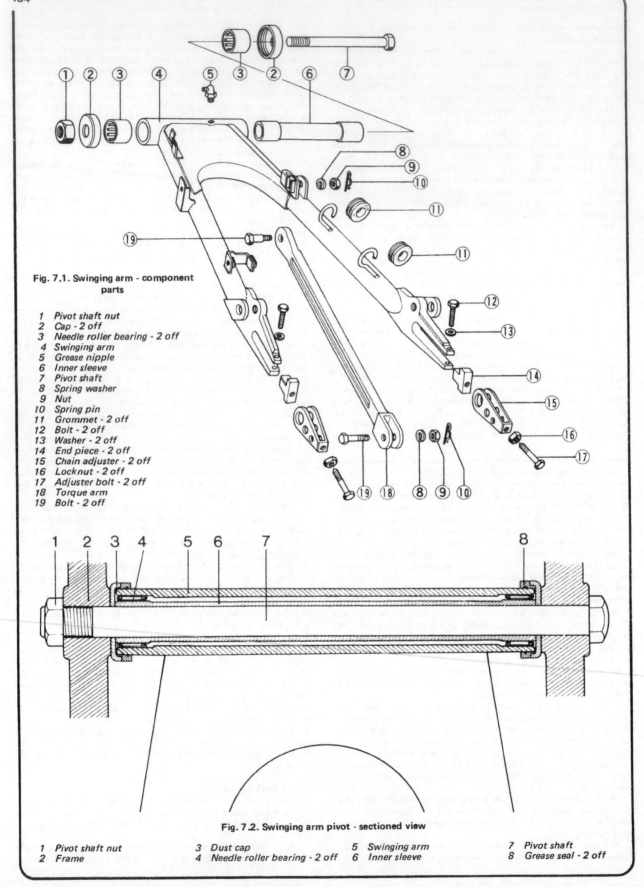

Fig. 7.1. Swinging arm - component parts

1 Pivot shaft nut
2 Cap - 2 off
3 Needle roller bearing - 2 off
4 Swinging arm
5 Grease nipple
6 Inner sleeve
7 Pivot shaft
8 Spring washer
9 Nut
10 Spring pin
11 Grommet - 2 off
12 Bolt - 2 off
13 Washer - 2 off
14 End piece - 2 off
15 Chain adjuster - 2 off
16 Locknut - 2 off
17 Adjuster bolt - 2 off
18 Torque arm
19 Bolt - 2 off

Fig. 7.2. Swinging arm pivot - sectioned view

1 Pivot shaft nut
2 Frame
3 Dust cap
4 Needle roller bearing - 2 off
5 Swinging arm
6 Inner sleeve
7 Pivot shaft
8 Grease seal - 2 off

torque arm from the brake plate by removing the single retaining nut. Detach the rear brake cable at the actuating lever.

12 Remove the wheel spindle nut completely, and withdraw the spindle. The wheel can now be drawn clear.

Disc brake models

13 Slacken the chain adjusters and wheel spindle nut, and remove the adjuster stops as described in paragraph 10.

14 Remove the spring pin, nut and bolt which retains the caliper to the torque arm. Push the wheel forward and disengage the chain from the rear wheel sprocket.

15 The wheel, together with the caliper, should be drawn backwards to clear the swinging arm. The caliper mounting lug can then be disengaged from the wheel spindle and tied clear of the swinging arm. Note that it may prove necessary to disconnect the hydraulic hose from the caliper. If this is the case, then the hose should be tied up at a higher level than that of the master cylinder, to avoid fluid loss. Take care not to drip the fluid on any paintwork. Unhook the hydraulic hose from the guides on the swinging arm.

All models

16 Remove the lower mounting bolts from the suspension units, and free the units from their mounting lugs.

17 Remove the pivot shaft nut, and withdraw the shaft. The swinging arm can now be pulled clear and disengaged from the chain.

18 Detach the left-hand footrest and the gearchange pedal, followed by the starter cover and gasket. Remove the engine sprocket cover after first removing its four retaining bolts.

19 Knock back the tabwasher, and remove the engine sprocket nut. The sprocket can be locked in position by bunching the chain against the casing.

20 Pull off the sprocket and disengage the chain. The new chain can be fitted by reversing the removal sequence. Ensure that the chain is adjusted to give 30 - 35mm (1.2 - 1.4 in) movement at the middle of the lower run. Tighten the wheel spindle to 10.0 - 14.0 kg m (72 - 101 ft lbs). If the hydraulic hose has been removed, bleed the rear brake system as described in Chapter 5, Section 16.

3 Hydraulic disc brakes: modifications

1 The increase in power output on the Z1000 models has been offset by an uprated braking system. In the UK, the single 296mm disc brake (still used on USA models) has been replaced by a twin 245mm configuration. A single 290mm disc has been fitted in place of the rear drum brake.

2 Front brake maintenance is similar, irrespective of the number of discs employed, and reference for this should be made to Chapter 5 for information.

4 Rear brake master cylinder: maintenance, removal and renovation

1 The rear brake master cylinder, and its integral reservoir, are located beneath the right-hand side panel. It is retained to the frame by two bolts. Movement from the brake pedal is conveyed mechanically to the master cylinder via an adjustable pushrod. Hydraulic pressure is transmitted from the master cylinder to the rear brake caliper by way of a flexible hydraulic hose.

2 The hydraulic fluid level is visible through the translucent sides of the reservoir, and should be checked weekly (or every 200 miles). Top up, if necessary, using only new hydraulic fluid which complies with DOT 3 standards. In the UK, the manufacturers recommend Castrol Girling Universal Brake and Clutch Fluid, or similar.

3 Avoid allowing the reservoir or its contents to become contaminated. It should be noted that hydraulic fluids are by nature hygroscopic; that is, they absorb water from the atmosphere. This in turn degrades the fluid specifications. To minimise water absorption, a rubber diaphragm is fitted beneath the vented reservoir cap. This rises and falls with the fluid level when the brakes are used, but prevents the ingress of damp air. The manufacturers recommend that the fluid should be changed annually. This is well worth doing, as it will flush out any impurities which may have accumulated, and which would otherwise tend to wear the cylinder and caliper components. In view of the small amount of fluid involved, it is not an expensive operation.

4 Leakage of the master cylinder, characterised by spongy and vague brake operation, and traces of fluid around the pushrod, will necessitate removal of the unit.

5 The master cylinder can be withdrawn after the banjo union on the brake pipe is detached, and the two mounting bolts removed. Take care not to spill any fluid on painted or plastic parts, as it will quickly attack these unless removed immediately.

6 The construction and operation of the rear master cylinder is very similar to that of the front unit, and the sequences described in Chapter 5, Section 14 can be applied. After the unit is refitted, bleed the system as described in Chapter 5, Section 16, to remove any air.

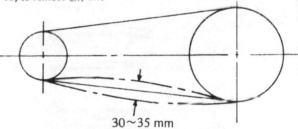

30~35 mm

Fig. 7.4. Chain tension adjustment
30 - 35 mm (1.2 - 1.4 in)

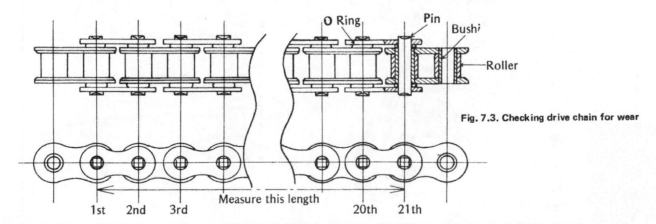

O Ring   Pin   Bush
Roller

Fig. 7.3. Checking drive chain for wear

Measure this length

1st 2nd 3rd   20th 21th

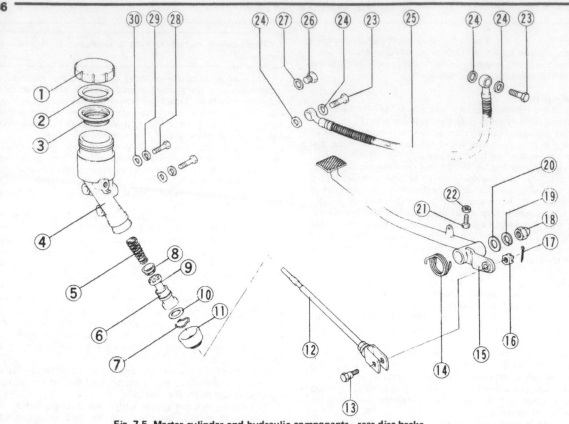

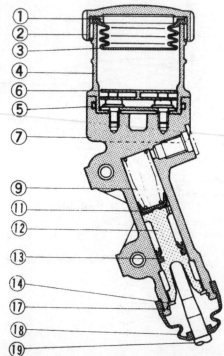

Fig. 7.5. Master cylinder and hydraulic components - rear disc brake

1 Reservoir cap	8 Primary seal	16 Castellated nut	24 Washer - 4 off
2 Washer	9 Piston	17 Split pin	25 Brake hose
3 Diaphragm	10 Washer	18 Domed nut	26 Plug
4 Master cylinder/reservoir body	11 Dust cover	19 Spring washer	27 Fibre washer
5 Return spring	12 Push rod	20 Washer	28 Mounting bolt - 2 off
6 Secondary seal	13 Shouldered bolt	21 Stop bolt	29 Spring washer - 2 off
7 Retainer	14 Pedal spring	22 Lock nut	30 Washer - 2 off
	15 Brake pedal	23 Banjo bolt - 2 off	

Fig. 7.6. Rear master cylinder - sectioned view

1 Washer
2 Cap
3 Diaphragm
4 Reservoir
5 O ring
6 Plate
7 Master cylinder body
9 Return spring
11 Seal
12 Piston
13 Secondary seal
14 Stop
17 Retainer
18 Dust cover
19 Push rod

5 Rear brake caliper: maintenance, removal and replacement

1   The rear disc pads can be checked visually for wear by removing the cover on the top of the caliper unit. There is a small step in the friction material of each pad, which indicates the maximum permissible wear limit. The pads must be renewed when this point is reached.

2   Remove the two spring clips which retain the guide pins in the caliper body. Pull out the guide pins, using a pair of pointed nosed pliers. Hold the anti-rattle springs in position as this is done.

3   Withdraw the pads and shims from the caliper body.

4   Replacement is a reversal of the above procedure. It should be noted that the caliper pistons will have to be pushed back to permit the insertion of the new pads. This is best done by placing a bleed tube over the end of the bleed nipple and slackening the latter to release the hydraulic pressure.

5   The system should be bled as described in Chapter 5, Section 16, after the pads are fitted.

6   Should attention to the caliper itself be necessary, it can be removed together with the rear wheel as described in Section 2, paragraphs 13 to 15. The pads should be removed as described in the foregoing paragraphs.

7   Remove the two socket screws which retain the left-hand caliper half and draw the halves apart. The two pistons are best removed by blowing them out with an air line or footpump. Wrap some rag around the caliper half during this process, to avoid spraying hydraulic fluid everywhere.

8   Examine the pistons and bores for signs of scoring or corrosion, either of which will cause seal wear and subsequent failure.

9   Fit new seals to each caliper half and reassemble in the reverse order of the dismantling sequence. Ensure that all the components are lubricated with clean hydraulic fluid during assembly. The hydraulic system must be bled when the caliper is refitted to the machine. See Chapter 5, Section 16.

10   Periodically inspect the hydraulic hose which connects the master cylinder and caliper, for signs of perishing or chafing. If in any way suspect, the hose must be renewed to obviate the risk of sudden, and dangerous, failure.

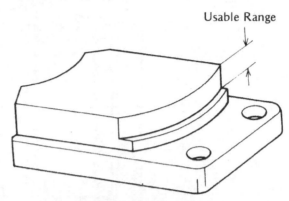

Rear Disc Brake Pad

Fig. 7.8. Step showing wear limit of rear pads

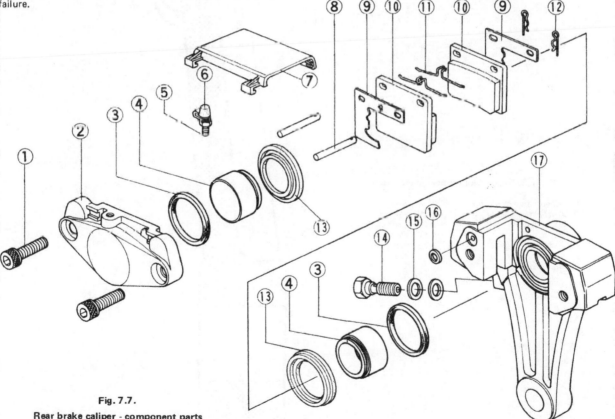

Fig. 7.7.

Rear brake caliper - component parts

1 Socket screw	6 Dust cap	10 Pad - 2 off	14 Banjo bolt
2 Left-hand caliper half	7 Cover	11 Anti-rattle spring - 2 off	15 Washer
3 Seal	8 Pin	12 Spring clip - 2 off	16 O ring
4 Piston	9 Shim	13 Dust seal	17 Caliper body
5 Bleed screw	10 Pad - 2 off		

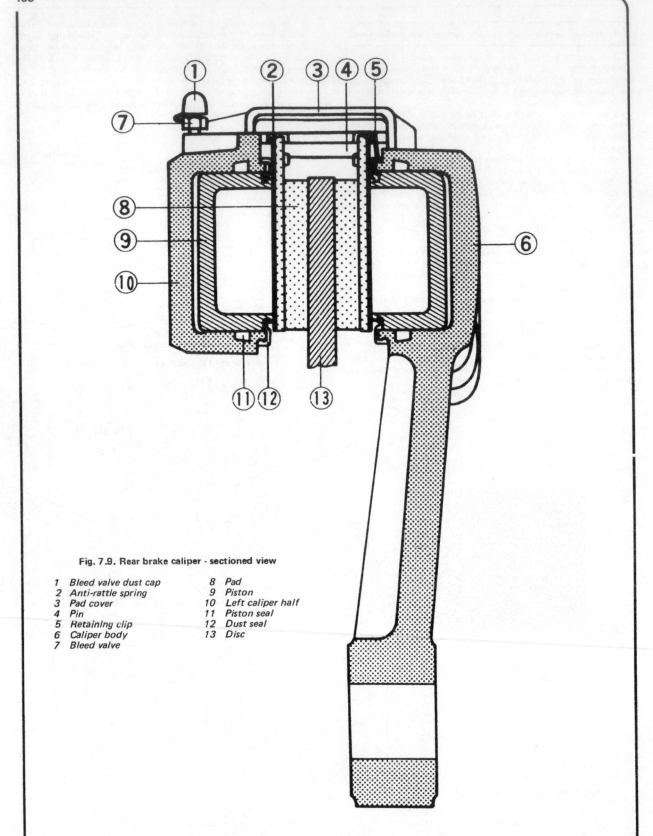

Fig. 7.9. Rear brake caliper - sectioned view

1	Bleed valve dust cap	8	Pad
2	Anti-rattle spring	9	Piston
3	Pad cover	10	Left caliper half
4	Pin	11	Piston seal
5	Retaining clip	12	Dust seal
6	Caliper body	13	Disc
7	Bleed valve		

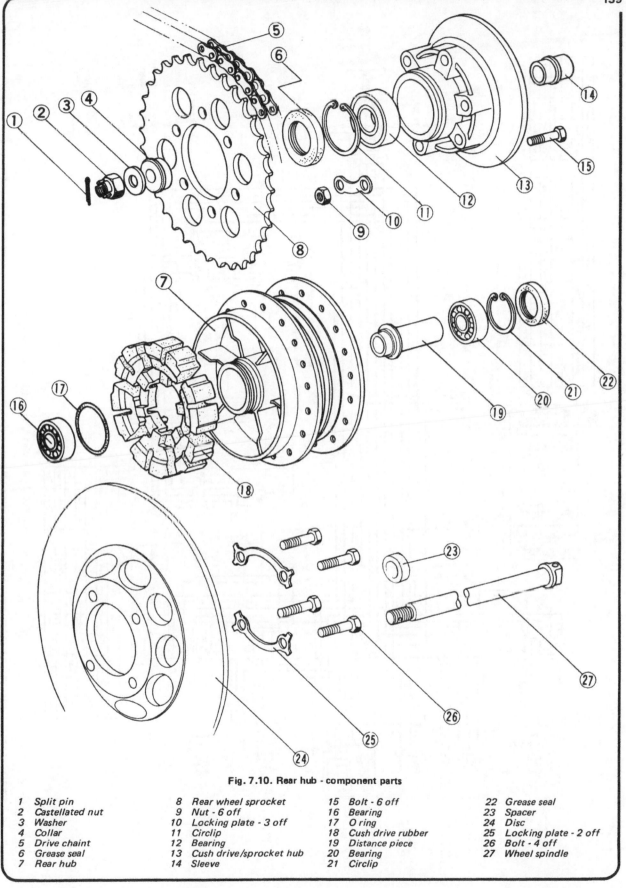

Fig. 7.10. Rear hub - component parts

1 Split pin	8 Rear wheel sprocket	15 Bolt - 6 off	22 Grease seal
2 Castellated nut	9 Nut - 6 off	16 Bearing	23 Spacer
3 Washer	10 Locking plate - 3 off	17 O ring	24 Disc
4 Collar	11 Circlip	18 Cush drive rubber	25 Locking plate - 2 off
5 Drive chaint	12 Bearing	19 Distance piece	26 Bolt - 4 off
6 Grease seal	13 Cush drive/sprocket hub	20 Bearing	27 Wheel spindle
7 Rear hub	14 Sleeve	21 Circlip	

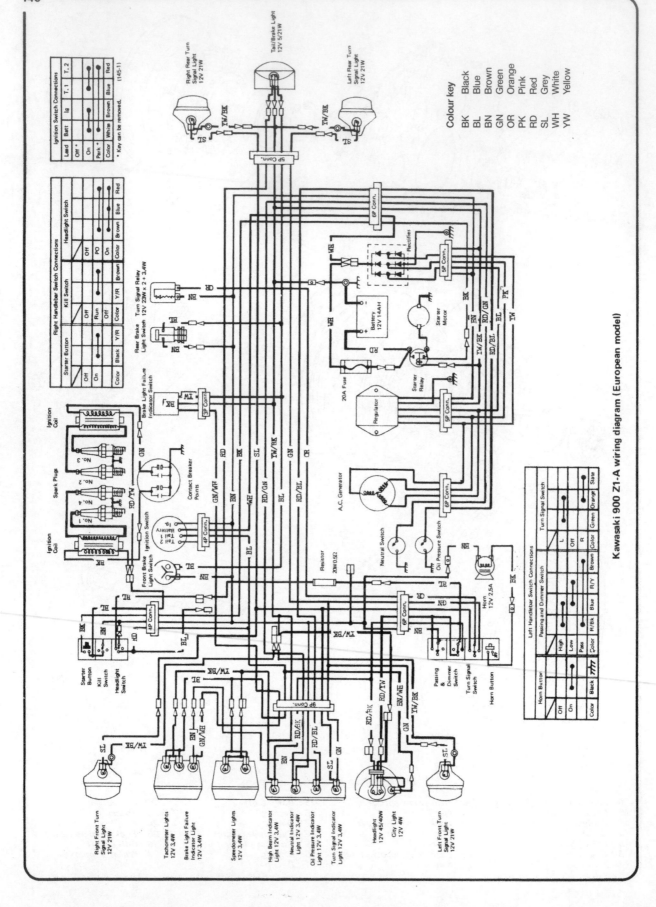

Kawasaki 900 Z1-A wiring diagram (European model)

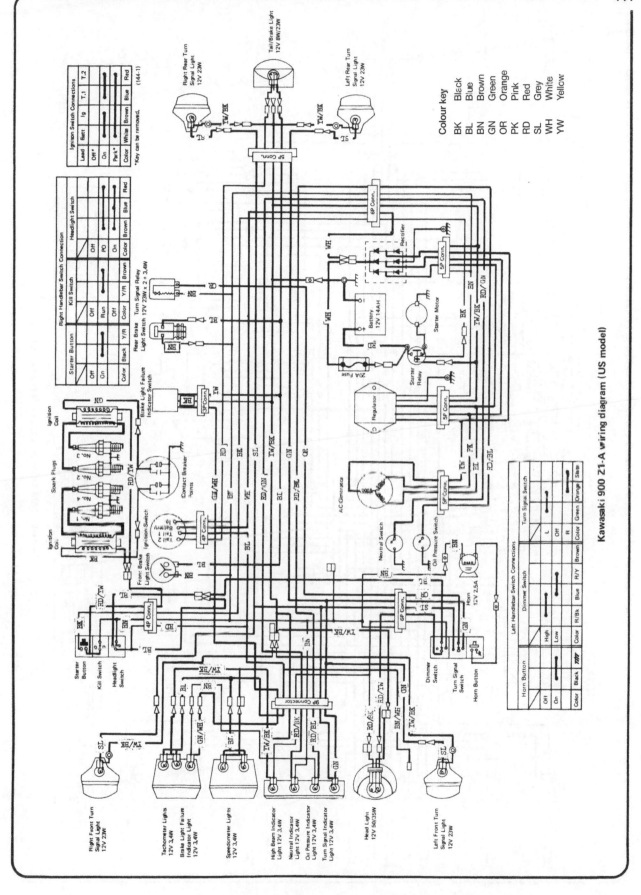

Kawasaki 900 Z1-A wiring diagram (US model)

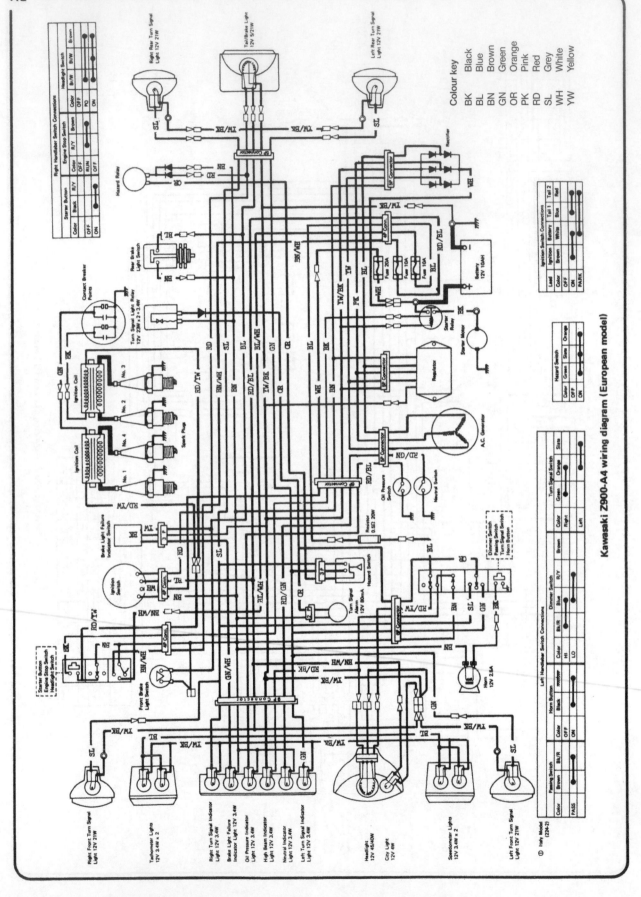

Kawasaki Z900-A4 wiring diagram (European model)

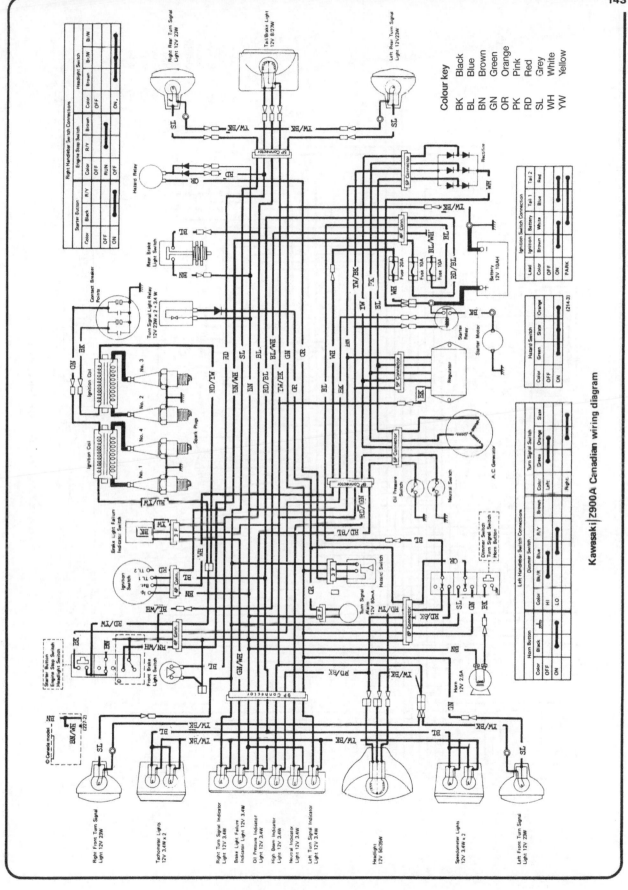

Kawasaki Z900A Canadian wiring diagram

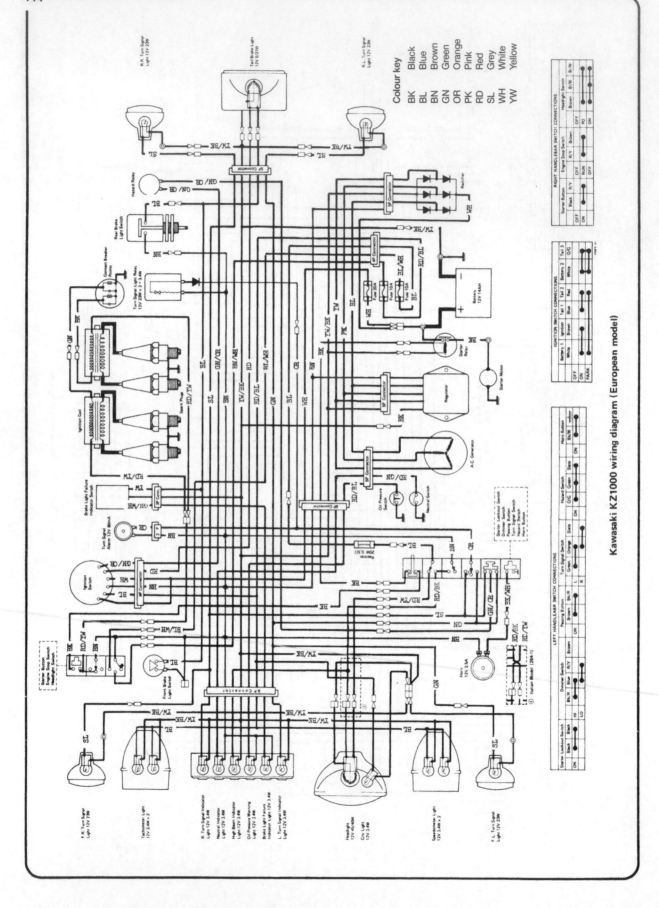

Kawasaki KZ1000 wiring diagram (European model)

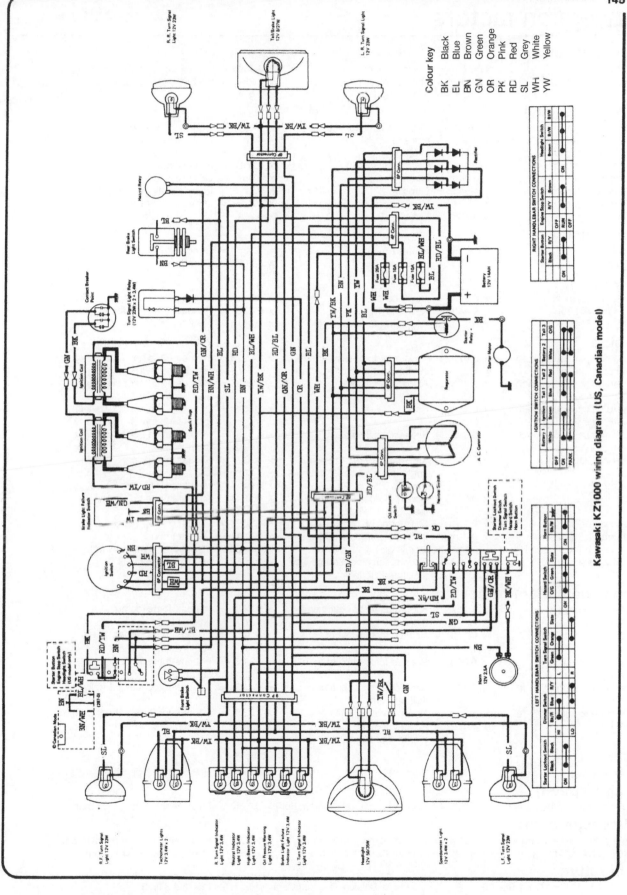

Kawasaki KZ1000 wiring diagram (US, Canadian model)

Colour key

BK	Black
EL	Blue
BN	Brown
GN	Green
OR	Orange
PK	Pink
RC	Red
SL	Grey
WH	White
YW	Yellow

Conversion factors

Length (distance)

Inches (in)	x 25.4	= Millimetres (mm)	x 0.0394	=	Inches (in)
Feet (ft)	x 0.305	= Metres (m)	x 3.281	=	Feet (ft)
Miles	x 1.609	= Kilometres (km)	x 0.621	=	Miles

Volume (capacity)

Cubic inches (cu in; in³)	x 16.387	= Cubic centimetres (cc; cm³)	x 0.061	=	Cubic inches (cu in; in³)
Imperial pints (Imp pt)	x 0.568	= Litres (l)	x 1.76	=	Imperial pints (Imp pt)
Imperial quarts (Imp qt)	x 1.137	= Litres (l)	x 0.88	=	Imperial quarts (Imp qt)
Imperial quarts (Imp qt)	x 1.201	= US quarts (US qt)	x 0.833	=	Imperial quarts (Imp qt)
US quarts (US qt)	x 0.946	= Litres (l)	x 1.057	=	US quarts (US qt)
Imperial gallons (Imp gal)	x 4.546	= Litres (l)	x 0.22	=	Imperial gallons (Imp gal)
Imperial gallons (Imp gal)	x 1.201	= US gallons (US gal)	x 0.833	=	Imperial gallons (Imp gal)
US gallons (US gal)	x 3.785	= Litres (l)	x 0.264	=	US gallons (US gal)

Mass (weight)

Ounces (oz)	x 28.35	= Grams (g)	x 0.035	=	Ounces (oz)
Pounds (lb)	x 0.454	= Kilograms (kg)	x 2.205	=	Pounds (lb)

Force

Ounces-force (ozf; oz)	x 0.278	= Newtons (N)	x 3.6	=	Ounces-force (ozf; oz)
Pounds-force (lbf; lb)	x 4.448	= Newtons (N)	x 0.225	=	Pounds-force (lbf; lb)
Newtons (N)	x 0.1	= Kilograms-force (kgf; kg)	x 9.81	=	Newtons (N)

Pressure

Pounds-force per square inch (psi; lbf/in²; lb/in²)	x 0.070	= Kilograms-force per square centimetre (kgf/cm²; kg/cm²)	x 14.223	=	Pounds-force per square inch (psi; lbf/in²; lb/in²)
Pounds-force per square inch (psi; lbf/in²; lb/in²)	x 0.068	= Atmospheres (atm)	x 14.696	=	Pounds-force per square inch (psi; lbf/in²; lb/in²)
Pounds-force per square inch (psi; lbf/in²; lb/in²)	x 0.069	= Bars	x 14.5	=	Pounds-force per square inch (psi; lbf/in²; lb/in²)
Pounds-force per square inch (psi; lbf/in²; lb/in²)	x 6.895	= Kilopascals (kPa)	x 0.145	=	Pounds-force per square inch (psi; lbf/in²; lb/in²)
Kilopascals (kPa)	x 0.01	= Kilograms-force per square centimetre (kgf/cm²; kg/cm²)	x 98.1	=	Kilopascals (kPa)
Millibar (mbar)	x 100	= Pascals (Pa)	x 0.01	=	Millibar (mbar)
Millibar (mbar)	x 0.0145	= Pounds-force per square inch (psi; lbf/in²; lb/in²)	x 68.947	=	Millibar (mbar)
Millibar (mbar)	x 0.75	= Millimetres of mercury (mmHg)	x 1.333	=	Millibar (mbar)
Millibar (mbar)	x 0.401	= Inches of water (inH₂O)	x 2.491	=	Millibar (mbar)
Millimetres of mercury (mmHg)	x 0.535	= Inches of water (inH₂O)	x 1.868	=	Millimetres of mercury (mmHg)
Inches of water (inH₂O)	x 0.036	= Pounds-force per square inch (psi; lbf/in²; lb/in²)	x 27.68	=	Inches of water (inH₂O)

Torque (moment of force)

Pounds-force inches (lbf in; lb in)	x 1.152	= Kilograms-force centimetre (kgf cm; kg cm)	x 0.868	=	Pounds-force inches (lbf in; lb in)
Pounds-force inches (lbf in; lb in)	x 0.113	= Newton metres (Nm)	x 8.85	=	Pounds-force inches (lbf in; lb in)
Pounds-force inches (lbf in; lb in)	x 0.083	= Pounds-force feet (lbf ft; lb ft)	x 12	=	Pounds-force inches (lbf in; lb in)
Pounds-force feet (lbf ft; lb ft)	x 0.138	= Kilograms-force metres (kgf m; kg m)	x 7.233	=	Pounds-force feet (lbf ft; lb ft)
Pounds-force feet (lbf ft; lb ft)	x 1.356	= Newton metres (Nm)	x 0.738	=	Pounds-force feet (lbf ft; lb ft)
Newton metres (Nm)	x 0.102	= Kilograms-force metres (kgf m; kg m)	x 9.804	=	Newton metres (Nm)

Power

Horsepower (hp)	x 745.7	= Watts (W)	x 0.0013	=	Horsepower (hp)

Velocity (speed)

Miles per hour (miles/hr; mph)	x 1.609	= Kilometres per hour (km/hr; kph)	x 0.621	=	Miles per hour (miles/hr; mph)

Fuel consumption*

Miles per gallon, Imperial (mpg)	x 0.354	= Kilometres per litre (km/l)	x 2.825	=	Miles per gallon, Imperial (mpg)
Miles per gallon, US (mpg)	x 0.425	= Kilometres per litre (km/l)	x 2.352	=	Miles per gallon, US (mpg)

Temperature

Degrees Fahrenheit = (°C x 1.8) + 32

Degrees Celsius (Degrees Centigrade; °C) = (°F - 32) x 0.56

It is common practice to convert from miles per gallon (mpg) to litres/100 kilometres (l/100km), where mpg x l/100 km = 282

Metric conversion tables

Inches	Decimals	Millimetres	Millimetres to Inches		Inches to Millimetres	
			mm	Inches	Inches	mm
1/64	0.015625	0.3969	0.01	0.00039	0.001	0.0254
1/32	0.03125	0.7937	0.02	0.00079	0.002	0.0508
3/64	0.046875	1.1906	0.03	0.00118	0.003	0.0762
1/16	0.0625	1.5875	0.04	0.00157	0.004	0.1016
5/64	0.078125	1.9844	0.05	0.00197	0.005	0.1270
3/32	0.09375	2.3812	0.06	0.00236	0.006	0.1524
7/64	0.109375	2.7781	0.07	0.00276	0.007	0.1778
1/8	0.125	3.1750	0.08	0.00315	0.008	0.2032
9/64	0.140625	3.5719	0.09	0.00354	0.009	0.2286
5/32	0.15625	3.9687	0.1	0.00394	0.01	0.254
11/64	0.171875	4.3656	0.2	0.00787	0.02	0.508
3/16	0.1875	4.7625	0.3	0.01181	0.03	0.762
13/64	0.203125	5.1594	0.4	0.01575	0.04	1.016
7/32	0.21875	5.5562	0.5	0.01969	0.05	1.270
15/64	0.234375	5.9531	0.6	0.02362	0.06	1.524
1/4	0.25	6.3500	0.7	0.02756	0.07	1.778
17/64	0.265625	6.7469	0.8	0.03150	0.08	2.032
9/32	0.28125	7.1437	0.9	0.03543	0.09	2.286
19/64	0.296875	7.5406	1	0.03947	0.1	2.54
5/16	0.3125	7.9375	2	0.07874	0.2	5.08
21/64	0.328125	8.3344	3	0.11811	0.3	7.62
11/32	0.34375	8.7312	4	0.15748	0.4	10.16
23/64	0.359375	9.1281	5	0.19685	0.5	12.70
3/8	0.375	9.5250	6	0.23622	0.6	15.24
25/64	0.390625	9.9219	7	0.27559	0.7	17.78
13/32	0.40625	10.3187	8	0.31496	0.8	20.32
27/64	0.421875	10.7156	9	0.35433	0.9	22.86
7/16	0.4375	11.1125	10	0.39370	1	25.4
29/64	0.453125	11.5094	11	0.43307	2	50.8
15/32	0.46875	11.9062	12	0.47244	3	76.2
31/64	0.484375	12.3031	13	0.51181	4	101.6
1/2	0.5	12.7000	14	0.55118	5	127.0
33/64	0.515625	13.0900	15	0.59055	6	152.4
17/32	0.53125	13.4937	16	0.62992	7	177.8
35/64	0.546875	13.8906	17	0.66929	8	203.2
9/16	0.5625	14.2875	18	0.70866	9	228.6
37/64	0.578125	14.6844	19	0.74803	10	254.0
19/32	0.59375	15.0812	20	0.78740	11	279.4
39/64	0.609375	15.4781	21	0.82677	12	304.8
5/8	0.625	15.8750	22	0.86614	13	330.2
41/64	0.640625	16.2719	23	0.90551	14	355.6
21/32	0.65625	16.6687	24	0.94488	15	381.0
43/64	0.671875	17.0656	25	0.98425	16	406.4
11/16	0.6875	17.4625	26	1.02362	17	431.8
45/64	0.703125	17.8594	27	1.06299	18	457.2
23/32	0.71875	18.2562	28	1.10236	19	482.6
47/64	0.734375	18.6531	29	1.14173	20	508.0
3/4	0.75	19.0500	30	1.18110	21	533.4
49/64	0.765625	19.4469	31	1.22047	22	558.8
25/32	0.78125	19.8437	32	1.25984	23	584.2
51/64	0.796875	20.2406	33	1.29921	24	609.6
13/16	0.8125	20.6375	34	1.33858	25	635.0
53/64	0.828125	21.0344	35	1.37795	26	660.4
27/32	0.84375	21.4312	36	1.41732	27	685.8
55/64	0.859375	21.8281	37	1.4567	28	711.2
7/8	0.875	22.2250	38	1.4961	29	736.6
57/64	0.890625	22.6219	39	1.5354	30	762.0
29/32	0.90625	23.0187	40	1.5748	31	787.4
59/64	0.921875	23.4156	41	1.6142	32	812.8
15/16	0.9375	23.8125	42	1.6535	33	838.2
61/64	0.953125	24.2094	43	1.6929	34	863.6
31/32	0.96875	24.6062	44	1.7323	35	889.0
63/64	0.984375	25.0031	45	1.7717	36	914.4

1   Imperial gallon = 8 Imp pints = 1.16 US gallons = 277.42 cu in = 4.5459 litres

1   US gallon = 4 US quarts = 0.862 Imp gallon = 231 cu in = 3.785 litres

1   Litre = 0.2199 Imp gallon = 0.2642 US gallon = 61.0253 cu in = 1000 cc

Miles to Kilometres		Kilometres to Miles	
1	1.61	1	0.62
2	3.22	2	1.24
3	4.83	3	1.86
4	6.44	4	2.49
5	8.05	5	3.11
6	9.66	6	3.73
7	11.27	7	4.35
8	12.88	8	4.97
9	14.48	9	5.59
10	16.09	10	6.21
20	32.19	20	12.43
30	48.28	30	18.64
40	64.37	40	24.85
50	80.47	50	31.07
60	96.56	60	37.28
70	112.65	70	43.50
80	128.75	80	49.71
90	144.84	90	55.92
100	160.93	100	62.14

lb f ft to Kg f m		Kg f m to lb f ft		lb f/in^2: Kg f/cm^2		Kg f/cm^2: lb f/in^2	
1	0.138	1	7.233	1	0.07	1	14.22
2	0.276	2	14.466	2	0.14	2	28.50
3	0.414	3	21.699	3	0.21	3	42.67
4	0.553	4	28.932	4	0.28	4	56.89
5	0.691	5	36.165	5	0.35	5	71.12
6	0.829	6	43.398	6	0.42	6	85.34
7	0.967	7	50.631	7	0.49	7	99.56
8	1.106	8	57.864	8	0.56	8	113.79
9	1.244	9	65.097	9	0.63	9	128.00
10	1.382	10	72.330	10	0.70	10	142.23
20	2.765	20	144.660	20	1.41	20	284.47
30	4.147	30	216.990	30	2.11	30	426.70

English/American terminology

Because this book has been written in England, British English component names, phrases and spellings have been used throughout. American English usage is quite often different and whereas normally no confusion should occur, a list of equivalent terminology is given below.

English	American	English	American
Air filter	Air cleaner	Number plate	License plate
Alignment (headlamp)	Aim	Output or layshaft	Countershaft
Allen screw/key	Socket screw/wrench	Panniers	Side cases
Anticlockwise	Counterclockwise	Paraffin	Kerosene
Bottom/top gear	Low/high gear	Petrol	Gasoline
Bottom/top yoke	Bottom/top triple clamp	Petrol/fuel tank	Gas tank
Bush	Bushing	Pinking	Pinging
Carburettor	Carburetor	Rear suspension unit	Rear shock absorber
Catch	Latch	Rocker cover	Valve cover
Circlip	Snap ring	Selector	Shifter
Clutch drum	Clutch housing	Self-locking pliers	Vise-grips
Dip switch	Dimmer switch	Side or parking lamp	Parking or auxiliary light
Disulphide	Disulfide	Side or prop stand	Kick stand
Dynamo	DC generator	Silencer	Muffler
Earth	Ground	Spanner	Wrench
End float	End play	Split pin	Cotter pin
Engineer's blue	Machinist's dye	Stanchion	Tube
Exhaust pipe	Header	Sulphuric	Sulfuric
Fault diagnosis	Trouble shooting	Sump	Oil pan
Float chamber	Float bowl	Swinging arm	Swingarm
Footrest	Footpeg	Tab washer	Lock washer
Fuel/petrol tap	Petcock	Top box	Trunk
Gaiter	Boot	Torch	Flashlight
Gearbox	Transmission	Two/four stroke	Two/four cycle
Gearchange	Shift	Tyre	Tire
Gudgeon pin	Wrist/piston pin	Valve collar	Valve retainer
Indicator	Turn signal	Valve collets	Valve cotters
Inlet	Intake	Vice	Vise
Input shaft or mainshaft	Mainshaft	Wheel spindle	Axle
Kickstart	Kickstarter	White spirit	Stoddard solvent
Lower leg	Slider	Windscreen	Windshield
Mudguard	Fender		

Index

A

About this manual - 2
Acknowledgements - 2
Adjustments:
    Brake - rear - 109
    Contact breaker - 85
    Headlamp beam height - 122
    Valve clearance - 55
Air cleaner - 78
Alternator - crankshaft - 120

B

Battery - 120
Bearings:
    Big end - 36
    Main - 36
    Steering head - 89
    Wheel - 106
Bleeding the hydraulic brake system - 115
Brake pedal - 97
Brakes:
    Falshing indicator - 126
    Headlamp - 122
    Specifications - 119
    Speedometer and tachometer - 126
    Stop and tail - 122

C

Cables: speedometer and tachometer drive - 101
Caliper: front disc brake - 112
Carburettors - 55, 69, 71, 75
Centre stand - 97
Chain: final drive - lubrication - 109
Cleaning the machine - 101
Clutch removal - 25
Clutch and rear chain oil pump - 40
Condensors - removal and replacement - 85
Contact breakers - adjustment - 85
Crankshaft - 42
Crankshaft alternator - checking output - 84
Cush drive - examination - 109
Cylinder block - examination - 37
Cylinder head - 18, 37, 42

D

Dimensions and weights - 4
Decarbonising - cylinder head - 37
Dualseat - removal - 99

E

Electrical system:
    Alternator - 120
    Battery - 120
    Fuse location - 122
    Handlebar switches - 126
    Headlamp - 122
    Horn location - 126
    Ignition and lighting switch - 126
    Indicator lamps - flashing - 126
    Starter motor - 122
    Stop and tail lamp - 122
    Voltage regulator - 121
    Wiring diagrams - 129, 130, 140, 141
Engine, clutch and gearbox:
    Decarbonising cylinder head - 37
    Dismantling - 18
    Examination and renovation - 36
    Lubricating system - 78
    Reassembly - general - 42
    Removal from frame - 13
    Refitting in frame - 55
    Starting and running a rebuilt unit - 56
Exhaust pipes removal - 13

F

Fault diagnosis:
    Clutch - 66
    Electrical system - 128
    Engine - 66
    Frame and forks - 101
    Fuel system - 81
    Gearbox - 67
    Ignition system - 87
    Lubrication system - 81
    Wheels and tyres - 96
Final drive chain - 109, 133
Final drive sprocket - 129
Flashing lamps - 126
Footrests - 97
Frame - examination and renovation - 97
Front wheel - 102 - 106
Fuel system:
    Air cleaner - 78
    Carburettors - 55, 69, 71, 75
    Petrol tank and tap - 69
    Tank capacity - 69
Fuse location - 122

G

Gearbox:
    Components - examination - 42
    Dismantling gear clusters - 33
    Rebuilding - 44

H

Handlebar switches - 126
Headlamp beam height adjustment - 122
Horn locations - 126
Hydraulic disc brake - modifications - 135

I

Ignition system:
    Coils - checking - 85
    Condensers - removal - 85
    Contact breaker adjustment - 85
    Spark plugs - 87
    Switch - 126
    Timing - 85, 87

L

Lamps - 122, 126
Legal obligations - 7, 118
Lighting switch - 126
Lubrication system - 78

M

Main bearings - 36
Maintenance - routine - 7
Master cylinders - 110, 135
Metric conversion tables - 142
Model dimensions and weights - 4

O

Oil filter - removal - 29
Oil pressure switch - 128
Oil seal - 37
Ordering spare parts - 5

P

Pedal - rear brake - 97, 109
Petrol tank and tap - 69
Pistons and rings - 18, 37
Prop stand - 97

Q

Quick glance routine maintenance - 9

R

Rear brake caliper - 137
Rear chain lubrication - 79
Rear cush drive - 109

Rear suspension units - 97
Rear wheel - 106
Rear wheel sprocket - 106
Recommended lubricants - 10
Ring - piston - 18 - 37
Routine maintenance - 7

S

Safety first! - 6
Seat dual - 99
Silicon rectifier - 120

Spark plugs - 87
Specifications:
    Bulbs - 119
    Brakes - 102
    Clutch - 12
    Electrical system - 119
    Engine - 11
    Fuel system - 68
    Ignition system - 82
    Wheels and tyres - 102
Speedometer - 99
Speedometer and tachometer drives - 101
Starter motor - 122
Statutory requirements - 7, 118
Steering head lock - 97
Stop and tail lamp - 122
Suspension units - rear - 97
Swinging arm fork - 97, 133

T

Tachometer - 101
Tappets - 39
Timing ignition - 85, 87
Tyres:
    Pressures - 9, 102
    Removal and replacement - 116
    Valves and dust caps - 116
Tyre changing sequence (colour) - 117

V

Valves, seats and guides - 39
Valve grinding - 39
Voltage regulator - testings - 121

W

Weights and dimensions - 4
Wheels:
    Balancing and alignment - 113
    Front - 102, 106
    Rear - 106
Wiring diagrams - 140, 141, 142, 143